Research for Development

The series Research for Development serves as a vehicle for the presentation and dissemination of complex research and multidisciplinary projects. The published work is dedicated to fostering a high degree of innovation and to the sophisticated demonstration of new techniques or methods.

The aim of the Research for Development series is to promote well-balanced sustainable growth. This might take the form of measurable social and economic outcomes, in addition to environmental benefits, or improved efficiency in the use of resources; it might also involve an original mix of intervention schemes.

Research for Development focuses on the following topics and disciplines:

Urban regeneration and infrastructure, Info-mobility, transport, and logistics, Environment and the land, Cultural heritage and landscape, Energy, Innovation in processes and technologies, Applications of chemistry, materials, and nanotechnologies, Material science and biotechnology solutions, Physics results and related applications and aerospace, Ongoing training and continuing education.

Fondazione Politecnico di Milano collaborates as a special co-partner in this series by suggesting themes and evaluating proposals for new volumes. Research for Development addresses researchers, advanced graduate students, and policy and decision-makers around the world in government, industry, and civil society.

THE SERIES IS INDEXED IN SCOPUS

Mirella Loda · Paola Abenante
Editors

Cultural Heritage and Development in Fragile Contexts

Learning from the Interventions
of International Cooperation in Afghanistan
and Neighboring Countries

 Springer

Editors
Mirella Loda
SAGAS—Storia, Archeologia, Geografia
Arte e Spettacolo
University of Florence
Florence, Italy

Paola Abenante
AICS, Office III Economic Opportunities
and Development
Italian Agency for Development
Cooperation
Rome, Italy

ISSN 2198-7300   ISSN 2198-7319 (electronic)
Research for Development
ISBN 978-3-031-54815-4   ISBN 978-3-031-54816-1 (eBook)
https://doi.org/10.1007/978-3-031-54816-1

This Springer imprint is published by the registered company Springer Nature Switzerland AG
The registered company address is: Gewerbestrasse 11, 6330 Cham, Switzerland

Paper in this product is recyclable.

Foreword by Amb. Teresa Castaldo

The Conference titled "Cultural Heritage in Fragile Contexts: Development Cooperation in Afghanistan and Neighboring Countries", was organized by the University of Florence, in collaboration with the Ministry of Foreign Affairs and International Cooperation, and the Italian Agency for Development Cooperation, and held on the 10th and 11th of November, 2022. The event provided an excellent opportunity to reflect on two lines of intervention that are traditional priorities for the Italian Cooperation: the protection of cultural heritage and the provision of support to countries in crisis, with a special focus on Afghanistan.

Italy has always played a leading role in the protection of cultural heritage on a global level. It has done so in awareness that such initiatives not only represent a major step towards inclusive and sustainable development but are also an essential tool to foster dialogue and mutual understanding.

The protection and enhancement of cultural heritage provides employment opportunities for workers in the sector and encourages work in the tourism and service sectors. At the same time, it contributes to creating an environment that is conducive to democratic development by fostering good governance and a sense of national belonging.

Over the last few decades, the Italian cooperation, which also relies on the expertise of a dense network of prestigious research centers, such as the Central Institute for Restoration and the Opificio delle Pietre Dure in Florence, as well as on the work of numerous universities, has carried out several initiatives in the field of heritage protection and promotion. Together, these are recognized internationally as best practices.

In fragile countries and crisis areas, where the situation is more complex, Italy can rely on the experience gathered over the years in countries such as Lebanon, Bosnia and Herzegovina, Iraq, Syria, Afghanistan, and most recently, Ukraine. In these contexts, our efforts are not only limited to the protection of the historical evidence of civilizations, but are part of a broader effort aimed at bolstering the resilience of local communities.

From time immemorial, the intentional destruction and misappropriation of cultural heritage has been one of the consequences of war and conflict: a way to demonstrate the victor's strength while weakening the identity of the defeated. In more recent times, the theft of cultural heritage works from conflict areas and their subsequent illegal commercialization has become a security issue not to be underestimated, as it represents a source of financing for criminal and terrorist organizations. This is why a collective response is vital, which in our case is articulated primarily through UNESCO and in coordination with our European Partners wherever possible.

Concerning the specific case of Afghanistan, our presence in the country, particularly in Bamiyan, can be termed *historical*. Unfortunately, the political situation that came about as a result of the Taliban's seizure of power forced us to effectively freeze bilateral initiatives.

Two notable exceptions have been the projects "Preservation and Promotion of the Bamiyan Valley through Culture-Oriented Sustainable Development" and "Local Community Empowerment and Preservation of Shahr-e Gholghola, the World Heritage site in Bamiyan", whose progress was the focus of the discussion at the Florence Conference.

It was in fact decided to prolong the two projects and respond favorably to UNESCO's request on the matter. The Organization had already adopted Decision No. 47 at the 212th Session of the Executive Council held in October 2021, a Decision which was also backed by Italy, in order to reaffirm its willingness to continue its work in Afghanistan. This was in line with the provisions of paragraph three of the United Nations Security Council Resolution 2593 of August 2021, which provided for the continuation of UNESCO's activities and programs in Afghanistan aimed at the protection and promotion of cultural heritage.

We hope that the continuation of these activities will also contribute to the best possible preservation and protection of the immense heritage of the Bamiyan area. It is a site that, not least by virtue of the still clearly visible scars left by the destruction of the Buddhas in 2001, will provide future generations with a concrete testimony of the role that cultural heritage can play in terms of respect for freedom of expression and the importance of intercultural and interreligious dialogue.

Amb. Teresa Castaldo
Director General for Development
Cooperation
Ministry of Foreign Affairs
and International Cooperation
Rome, Italy

Foreword by Leonardo Carmenati

This volume results from an AICS-funded cooperation action in Afghanistan, carried out by the University of Florence, titled *"Redevelopment of Informal Settlements and Location of Population Flows in Bamiyan and Herat"*. First started in 2019, the project addresses urban planning associated with the management and enhancement of cultural and natural heritage, and it concerns itself in particular with Herat and Bamiyan; two cities rich in culture, and in the case of the Bamiyan Valley, a World Heritage Site.

The initiative emerged amidst a period of historic significance: when interventions by the Italian Cooperation in Afghanistan were providing support to the program "Afghanistan National Development and Peace Framework 2017–2021" and the related National Priority Plans (NPPs). This support was directed at the efforts of the then-Afghan authorities, for the sustainable development of progressive urbanization, across areas rich in cultural heritage. The project complements two other UNESCO-run initiatives funded by the Italian Cooperation in 2016 and 2018, aimed at safeguarding and enhancing the Bamiyan World Heritage Site, by training local staff in the management of the site, strengthening community ties, and creating employment and training opportunities in the cultural sector.

Following the events that occurred in Afghanistan in August 2021, and given the country's shifting political landscape, these projects have had to reassess their objectives, in order to capitalize on actions carried out on the ground until 2021, without allowing progress already made, to go to waste.

Regarding developmental interventions in the cultural field, Italian-funded projects have been the first to resume operations in the country. This has again been done under the Transitional Engagement Framework, promoted by the United Nations and which aimed at directing and coordinating actions in Afghanistan in 2022, affirming the need to act without interacting with the country's de facto authorities. This Transitional Framework lays out three priority areas of engagement: humanitarian assistance, support for essential services, and the promotion of investments in social and community support services. The projects financed by the Italian Cooperation are specifically geared towards the third commitment: mitigating social and economic fragility among the population. The alleviation of fragility passes, on the one hand, through the symbolic power of a World Heritage Site of Bamiyan, which is able to keep open a channel of communication between the local community and the rest of the world. In a sense, this allows the population to come out of isolation and stay connected with what happens and what is discussed outside the geographical and political boundaries of contemporary Afghanistan. On the other hand, initiatives in Bamiyan concretely support the area's local communities through the planning of activities which leave wide space for cash-for-work, thus directly funding the local population in ways that benefit the community itself. Examples of this would be the redevelopment of the urban area surrounding the site and inhabited by the communities themselves, the conservation and securing of the archaeological site, and the support to women through the development of activities related to the infrastructure of the archaeological site.

At the same time, the University of Florence, in coordination with the Culture and Development unit of AICS Office III, has strengthened the research and dissemination component of the project financed by AICS. This is with the aim of capitalizing on the results of extensive research, which raises the issue of the social and economic impact of cooperation interventions in the cultural heritage sector. The conference from which this volume has derived was one of such dissemination actions organized for this very purpose.

Leonardo Carmenati
Deputy Technical Director
AICS—Agenzia Italiana per la
Cooperazione allo Sviluppo
Rome, Italy

Acknowledgements

In bringing this book to a close, we would like to thank the Italian Agency for Development Cooperation (AICS) and the Italian Cooperation, for making possible the International Conference *Cultural Heritage in Fragile Contexts—Development Cooperation in Afghanistan and Neighbouring Countries* (Florence, November 10–11, 2022), which offered the opportunity for international discussion on an issue central to development cooperation, and which inspired the reflections put forward in these pages.

We would also like to thank the Afghan students at the University of Florence for their commitment and enthusiasm in helping sort through data and information regarding their country, in a context of great complexity and uncertainty.

Finally, we would like to thank Adelina Sultan for her support in cartographic and GIS processing; Angeliki Coconi for her support in translating and editing the texts; and Camilla Froio for her help in organizational and administrative matters.

This book has been financed by the Italian Agency for Development Cooperation and by the University of Florence. Any proceeds from this volume will be devolved to the safeguarding of cultural heritage.

Contents

Suggestions for Field Work

Introduction

Mirella Loda⊙ and Paola Abenante

Abstract This volume offers contributions for reflection on cooperation actions in fragile contexts, starting from the discussions that took place during the international conference "Cultural Heritage in Fragile Contexts—Development Cooperation in Afghanistan and Neighboring Countries", held in Florence from the 10th to the 11th of November 2022. The event offered an opportunity for international discussion between researchers and international operators, as well as a chance to reflect on key relationships between the protection of cultural heritage and the generation of sustainable development. More specifically, focus was placed on the ways in which cultural protection actions can help mitigate social and economic fragilities in intervention areas, contributing to their sustainable development. The geographical focus of the conference, and of this volume, are the Middle and Central Asian areas, as these are areas where situations of fragility are recurrent, both in the socio-cultural sense—just think of the religious tensions between the Shiite and Sunni populations—as well as in political-military terms—the cases of Afghanistan and Syria are exemplary in this regard.

The authors have contributed to the present essay as follows: Mirella Loda, Sections 2 and 3; Paola Abenante, Section 1; Section 4 has been written by both authors.

M. Loda (✉)
Department of History, Archaeology, Geography, Fine and Performing Arts (SAGAS) and Laboratory for Social Geography (LaGeS), University of Florence, Via San Gallo 10, 50129 Firenze, Italy
e-mail: mirella.loda@unifi.it

P. Abenante
Italian Agency for Development Cooperation, Via Cantalupo in Sabina 29, 00191 Rome, Italy
e-mail: paola.abenante@aics.gov.it

1 Fragility

Fragility is indeed a complex and much debated concept in development approaches.[1] For the purposes of this volume, OECD's definition of fragility comes in handy to set the conceptual background for our reflections. OECD determines fragility through a multidimensional conceptual framework that takes into account not only the political and security dimensions of a context, but also the economic, societal, environmental dimensions and, more recently, the human dimension. At the same time, OECD understands fragility as a dynamic condition that varies in intensity and nature, according to the ability of states, systems and local communities to absorb and cope with the different types of risks to which they are exposed. A context is all the more fragile the less the preparedness, response and coping capacity of states, systems and communities (OECD 2022).

Such a framework marks a turning point with regard to previous ways of understanding fragility, for a number of reasons, in the perspective of developing a long-term and comprehensive approach to fragile contexts in line with the triple humanitarian-development-peace nexus (OECD 2022).[2] Suffice it here to highlight one relevant change introduced by the framework that is the overcoming of a state-centered understanding of fragility and the recognition of civil society institutions—*system and community*—beyond the *state*, as essential in preventing, absorbing and withstanding recurrent shocks and protracted crises. Indeed, in contexts of fragility, where state governance is weak, it is all the more necessary to promote a people-centered development approach that supports the coping capacities not only of formal institutions but also of society at large, by recognizing and reinforcing informal social institutions and community level forms of organization. Accordingly, the task of development cooperation is to support fragile contexts also by identifying and leveraging endogenous socio-cultural resources that are at the root of community life, such as locally established social practices, shared meanings and values, hence empowering local forms of civil society. Indeed, through such resources, communities and societies nurture cohesion and resilience, that is to say, the capacity to get together and to respond to difficulties.

The OECD Fragility Framework does not specify the role of culture in the development of societies' coping capacities. Yet UNESCO and other expert bodies have consistently advocated the positive social and economic impacts that actions supporting culture and cultural heritage protection have on post-conflict recovery, crisis-response and in peacebuilding interventions. It seems useful for the purposes

---

[1] Several attempts have been made to design a comprehensive and effective methodology to measure and confront it, see (OECD 2022) for an overview on approaches to fragility.

[2] By taking into account six possible dimensions inducing fragility, the OECD framework nuances the divide between fragile and non-fragile states. On the other hand, it implies that any context can carry a degree of fragility. Moreover, by measuring both risks and coping capacities, the framework assesses also weaknesses and mitigation measures. This allows for an approach, in line with the humanitarian-development-peace nexus, adopted by the Development Assistance Committee in 2019, with the aim of 'shifting from delivering humanitarian assistance to ending need' (OECD 2022).

of this volume to bring forth an underlying understanding of how culture counts in confronting fragility and in laying the foundations of a people-centered development approach.

Building on the definition of culture in the UNESCO 2001 *Universal Declaration on Cultural Diversity*, one can conceive culture as both a social process of making meaning and a product of human expression: "a set of distinctive spiritual, material, intellectual and emotional features of society or a social group, and that encompasses, in addition to art and literature, lifestyles, ways of living together, value systems, traditions and beliefs" (UNESCO 2001). Culture can be understood as the lens through which individuals and communities come to know, interpret and express reality. Heritage, cultural imaginaries and memories are the matrix of communities' shared meanings and ways of life, the bond that keeps communities together and grounds common aspirations for the future. Therefore, there is a need to recognize culture as a resource inherently belonging to people, which helps strengthen livelihoods and social cohesion, increasing communities' capacity to absorb and to cope with difficulties and risks.

During the latest *Mondiacult* conference, which was held in Mexico in September 2022 and aimed at the promotion of culture within public policy as well as its integration as a stand-alone goal in the 2030 Agenda, UNESCO drafted a declaration defining culture as a *global public good*. Borrowing the words of Ottone Ramirez, UNESCO's Assistant Director General for Culture, "Culture is what defines us. It is the common thread between our past, present and future. It is an inexhaustible and renewable resource, which adapts to changing contexts and which drives our capacity to imagine, create, innovate" (UNESCO 2022).[3]

The need to protect and leverage culture as a global public good is all the more important in fragile contexts, where resources are scarce and difficult to mobilize. Indeed, as a process of making meaning, culture is a resource that inherently belongs to people, communities and societies, and that may not be alienated. Moreover, as a product of human expression, culture is a universally available resource, never scarce. Finally, as a set of shared imaginaries and memories, linking the past to the present and the future, it is the tool through which communities build and continuously re-adapt their identity.

At the same time, as the contributions to this volume show, culture or, rather, *cultures* are complex, diverse and continuously changing. Indeed, processes of making meaning are context-specific, ever evolving and readapting when faced with change, never coherent but rather dependent on power relations and different interests among groups and individuals within the society. An analytical approach to

---

[3] Such attention to culture as a resource for development digs its roots back into the 'sustainability' shift in development policies and practices for which, in the last decades of the twentieth century, the 'economic growth model' of development has been replaced by the three pillars of the sustainability model—social equity, environmental protection and economic viability—alongside the focus on human development. Since then a reflection on the role of culture in development gained momentum, grounding progressively culture into policy documents. However such reinstating of the importance of culture not often has been mirrored in development actions, and several development actors have been working on research and advocacy of culture in development actions.

contexts of cooperation interventions that is able to recognize and address such complexity is necessary in order to properly leverage the power of culture in sustainable development.

It is by bearing this in mind that we should read through the volume: conceptual tools such as *memory, landscape, identity, diversity, livelihoods and living traditions* are placed at the forefront, attesting to an understanding of culture—or rather *cultures*—and cultural heritage, in terms of continuously evolving, diverse and contextual expressions of human society, rather than in terms of unalterable artifacts. These concepts all highlight three dimensions of cultural heritage that are dealt with: heritage as contextual; heritage as relational and heritage as dynamic. Heritage is the contextual expression of a diversity of cultures and territories. Heritage is relational in as much as it exists when recognized as such by one or more parties, be it UNESCO or the local stakeholders, and it carries different nuances and values depending on who it is that is recognizing it. Heritage is dynamic because it carries the specific history of the community it belongs to and, at the same time, because it continuously renews itself in relation to contemporary understandings and future aspirations.The analytical approaches to such complex understanding of cultural heritage that emerge in the following chapters aim to contribute to the effort of translating policy recommendations on the role of culture into concrete development practices where cultural heritage is not just about monuments in need of protection but is, rather, a multifaceted resource belonging to the people.

2 Authenticity

What has led to such an understanding of cultural heritage as belonging to humankind and at the same time as highly intricate in the range of its manifestations, goes all the way back to the 1972 UNESCO *Convention concerning the Protection of the World Cultural and Natural Heritage* (UNESCO 1972). Since then, the idea that there are forms of cultural and natural heritage not ascribable to a limited cultural sphere but of universal interest and value as a product of human expression, has taken shape. At the same time a complex system of actions, carried out by public and private bodies, aimed at the identification and protection of this heritage, has begun to unfold at an international scale.

As is well understood, the path that opened up following the 1972 *Convention* was not linear. The initial notion of universal, cultural or natural property has been subjected to much criticism over the years. In particular, its notable conformity to conceptual schemes specific to Western culture, and its inadequacy when it comes to representing heritages that are the expression of other cultural contexts, has been emphasized.

Indeed, the documents that initially laid out the guiding criteria for the protection of cultural heritage stressed the urgency in preserving cultural sites in their material authenticity and integrity. Material authenticity, however, is hard to hold onto in cultures whose architecture is built with mud bricks, as is the case in Central Asian areas; or where temples are traditionally built using wood and ritually burnt down at regular intervals to be reconstructed identically, as is the case in East Asian areas; or, where heritage consists largely of gardens, like in East Asian or Persian countries.

Therefore, from the 1990s onwards, UNESCO endeavored to correct the original approach to Cultural Heritage and to re-introduce discussions on that most crucial concept, authenticity. The most important forerunner of these revisions is the Nara Document on Authenticity issued in 1994 by ICOMOS/Japan. Its §11 reads: "All judgements about values attributed to cultural properties as well as the credibility of related information sources may differ from culture to culture, and even within the same culture. It is thus not possible to base judgements of values and authenticity on fixed criteria. On the contrary, the respect due to all cultures requires that heritage properties must be considered and judged within the cultural contexts to which they belong" (ICOMOS/Japan 1994).

ICOMOS' *Nara + 20 Conference On Heritage Practices, Cultural Values and the Concept of Authenticity*, 2014, defined this revised concept of authenticity at greater length: It is "a culturally contingent quality associated with a heritage place that is perceived to communicate credibly cultural value, that is recognized as a meaningful expression of an evolving cultural expression, and/or evokes among individuals the social and emotional resonance of group identity" (UNESCO 2023).

The official documents issued by UNESCO itself continue along this line of thought. UNESCO's latest *Operational Guidelines for the Implementation of the World Heritage Convention* (2023) phrase the problem of authenticity quite cautiously (§82): "Properties may be understood to meet the conditions of authenticity if their cultural values … are truthfully and credibly expressed" (ICOMOS 2014). Authenticity of cultural sites (and also integrity of natural sites) here becomes a problem of "expression", which must at least be "credible", if not "truthful"; in the final analysis, it becomes an issue of rhetorical self-representation.

A further attempt to address the initial Eurocentric nature of UNESCO's approach consisted in the introduction of "Intangible Heritage", achieved in 2003 under the directorate of Koïchiro Matsuura (UNESCO 2003). The concept of intangible heritage was aimed at overcoming the monument-oriented character of the World Heritage List and to integrate social communities, with their traditions, customs and crafts, into the realm of items deserving of not only protection but also recognition of their outstanding universal value. However, the distinction between tangible and intangible, as well as material and immaterial heritage has been criticized too, especially by the School of *Critical Heritage Studies* (Smith 2006) on the basis that heritage is not a *thing* but a usage, embedded in a specific historical and social context, and therefore necessarily intangible.

Finally, a "bottom-up" approach to the recognition of cultural heritage has gained ground, especially with the 2001 UNESCO *Universal Declaration on Cultural Diversity* (UNESCO 2001) and the *Faro Convention on the Value of Cultural Heritage for Society* held by the European Council in 2005 (European Council 2005). Said approach aims at recognizing cultural diversity as humanity's common cultural patrimony and cultural heritage as a product of societies/communities' expression and aspirations. This understanding of heritage has widened the fragmentation and diversification of the concept itself, bringing it to encompass highly heterogeneous categories of assets.

Taken together, however, these critical reflections and relevant adjustments to official documents have allowed for the notion of heritage to adapt to different cultural contexts, shifting emphasis from the idea of an asset as a concrete product, to that of an asset as a product and a process of continuous (re)-signification in relation to the cultural context of origin. Today, UNESCO World Heritage counts as many as 1199 sites spread over 168 countries, and around 500 items are included on the intangible heritage list, while a highly articulated protocol defines the criteria for World Heritage recognition and even the actions necessary to ensure protection of such assets.

The constant tension between the universal approach underlying the World Heritage List and the extraordinary multiplicity of existing cultural forms and creations on a global scale is also reflected at the procedural level. The requirement for sites to have a management plan—a protection and management system that ensures the maintenance and improvement of the Outstanding Universal Value through a coordinated apparatus of legislative, regulatory and managerial measures, using participatory methods—represents an interesting balancing point between the attempt to formulate a universal protocol for protective actions on the one hand, and the need to adapt them to the highly fluid peculiarities of the local contexts in which sites are located, on the other. In this regard, it is telling that no single model has been furnished for the preparation of the Management Plan.[4]

Overall, therefore, the multiplicity of cultural heritage contexts and—in some ways—their irreducibility, is highlighted. But, at the same time, a transcultural (universal) focus on the importance and protection of cultural heritage can now be thought to have been established. Also, there is at this point an awareness of the fact that a nexus exists between heritage on the one hand and other important territorial context dimensions on the other; especially a nexus between heritage protection, economic development and sustainability.

3 Development and Sustainability

The idea that "heritage is more than just monuments" (ICOMOS 2021, p. 12) has been firmly established in the UNESCO documents as well as its advisory bodies over the last decade.[5] Considerations on the connection between heritage preservation

---

[4] A helpful set of guidelines for the creation of a Management Plan has been jointly put together by ICCROM, ICOMOS, IUCN and UNESCO in 2013 (ICCROM, ICOMOS, IUCN, UNESCO 2013).

[5] We may recall some of the most relevant policy documents adopted in recent years, including the 40th anniversary of the *World Heritage Convention* in Kyoto 2012 under the title: *World Heritage and Sustainable Development: the Role of Local Communities*; the *Hangzhou Declaration Placing Culture at the Heart of Sustainable Development Policies* (2013); the *Florence Declaration Culture, Creativity and Sustainable Development. Research, Innovation, Opportunities* (2014); the 4th UN Resolution on *Culture and Sustainable Development* (2014) the *Policy Document for the Integration*

and the socio-economic development of the community involved were stoked by the problems posed by safeguarding cultural heritage in urban contexts and fixed in the *Historic Urban Landscape* (HUL) approach of 2011: "The historic urban landscape approach is aimed at preserving the quality of the human environment, enhancing the productive and sustainable use of urban spaces, while recognizing their dynamic character, and promoting social and functional diversity. It integrates the goals of urban heritage conservation and those of social and economic development. It is rooted in a balanced and sustainable relationship between the urban and natural environment, between the needs of present and future generations and the legacy from the past" (UNESCO 2011, §11). Special attention is paid to "preserving the quality of the human environment (...) while recognizing [its] dynamic character and promoting social and functional diversity".

The road was paved by ICOMOS' *Paris Declaration on heritage as a driver of development*, adopted on the 1st of December 2011,[6] and featuring the slogan "No Past, No Future" (ICOMOS 2011). The debates over the following decade showed that the opposite is also true, meaning that there cannot be a past without a debate about the future. The main scope of the *World Heritage Convention*, of course, lies in the safeguarding of the past cultural or natural heritage; in the reflections regarding the management of such heritage sites, however, development requirements will necessarily come up.

The management of heritage sites is inevitably a management of change. The joint ICCROM, ICOMOS, IUCN and UNESCO handbook on *Managing World Heritage* (2013) assumes the broader concept of heritage management introduced by the *Paris Declaration*: "The expanding concept of heritage and the increased importance given to how heritage places relate to their surroundings mark an important shift in thinking. Heritage places cannot be protected in isolation or as museum pieces, isolated from land-use planning considerations. Nor can they be separated from development activities, isolated from social changes that are occurring, or separated from the concerns of the communities" (ICCROM, ICOMOS, IUCN, UNESCO 2013, p. 12). ICCROM here expresses a clear understanding that heritage preservation policies must be part and parcel of the general territorial planning tools of a protected territory, which cannot neglect the development necessities of the local community. The authors further state that heritage can no longer be "confined to the role of passive conservation of the past but should instead provide the tools and framework to help shape, delineate and drive the development of tomorrow's societies" (ICCROM, ICOMOS,

---

of a Sustainable Development Perspective into the Processes of the World Heritage Convention (UNESCO 2015) and the *Hangzou Declaration Culture for Sustainable Cities* (2015).

[6] The *Paris Declaration* refers back to a previous ICOMOS conference held in 1978 at Moscow and Suzdal in Russia on the topic of *The Protection of Historical Cities and Historical Quarters in the Framework of Urban Development*. One of the sub-themes of the Russian conference was *Historical Monuments as a Support to Economic and Social Development*. ICOMOS' Paris Declaration, three decades later, concentrates on Urban Development too; point 4, however, puts forth a more general claim: "Place heritage at the heart of overall development strategies, setting goals for economic and social benefits to ensure that the development of heritage rewards local communities in terms of employment, the flow of finance and well-being".

IUCN, UNESCO 2013, p. 19). Consequently, the authors conclude, the management of heritage sites "would no longer be left in the hands of heritage experts but discussed among many counterparts" (ICCROM, ICOMOS, IUCN, UNESCO 2013, p. 21). We will deal more extensively with this point in the final chapter.

A further step in viewing heritage protection as a key component of territorial development was taken with the *Sustainable Development Goals* of the *Agenda 2030* adopted by the UN General Assembly in 2015. UNESCO integrated the concept of sustainability as part of its *Policy for the Integration of a Sustainable Development Perspective into the Processes of the World Heritage Convention* (WHC Resolution 20 GA 13, 2015) (UNESCO 2015). With reference to UNESCO's *World Heritage Convention* (§5), the *Operational Guidelines* and the 2002 *Budapest Declaration*, the World Heritage Committee demands that "States Parties should recognize, by appropriate means, that World Heritage conservation and management strategies that incorporate a sustainable development perspective embrace not only the protection of the Outstanding Universal Value, but also the wellbeing of present and future generations" (#6) and therefore "States Parties should integrate conservation and management approaches for World Heritage properties within their larger regional planning frameworks" (#10).

The sustainability concept established in the UN *Agenda 2030* will remain a benchmark in all further considerations of the nexus between heritage preservation and local development.[7]

It is precisely the understanding of such a nexus that calls into question the role of development cooperation activities, in their greater or lesser capacity, when it comes to planning heritage preservation actions that are.

(a) based on an adequate reading of the socio-cultural context and the dynamics of interaction between its various components, and

(b) capable of orienting the context toward objectives of social balance and sustainable development.

Therefore, the general commitment, which the Cooperative system is called upon, is extremely complex.

The complexity and subtlety of the task increase exponentially in territorial realities that are characterized by fragility, whether of a social, environmental or warlike nature, and where development cooperation action is forced to intervene in conditions of urgency, danger, tension or social conflict.

How is one to proceed in such cases? What objectives should be pursued? What forms of heritage should be protected as a priority? How could the concept of authenticity be adapted? These and other questions arise alongside interventions in fragile realities, albeit often in an implicit form.

---

[7] As ICOMOS stated in its 2021 report on *Heritage and Sustainable Development Goals* (ICOMOS 2021, p. 15) the formula used at point 11.4 of UN's sustainable goals concerning "Sustainable Cities and Communities" and highlighting the need to "strengthen efforts to protect and safeguard the world's cultural and natural heritage" had been, in a certain sense, originally put forth in the 2011 *Recommendation on the Historic Urban Landscape* (UNESCO 2011).

4   The Structure of the Book

The contributions offered in the following pages have been the result of the issues and cases discussed at the conference.

A large part of the volume, as well as of the conference, focused on Afghanistan, whose cultural heritage the international community has been very committed to protecting. For Afghanistan, in fact, the incorporation of heritage protection policies into a development perspective appears particularly urgent, as was highlighted at the 2016 Rome Conference *Cultural Heritage & Development Initiatives*, which was organized by the Islamic Republic of Afghanistan, World Bank and UNESCO (2016).[8]

After the abrupt interruption of all field activities that followed the establishment of the Emirate in August 2021, the Florentine event provided the first opportunity to take stock of the situation, but also, and above all, to discuss the risks associated with the new context, whether a possible re-engagement in the region is advisable, and the ways to go about it.

The first part of the volume brings together a series of studies on the cultural heritage protection and enhancement projects carried out in the Bamiyan Valley, one of the two UNESCO sites in Afghanistan upon which, since the fall of the first Taliban regime and its inclusion on the World Heritage List in danger (2003), the attention of international cooperation and donor countries has been largely focused: mainly Italy, France, Germany, Japan and South Korea.

The first chapter offers a broad overview of the acquisitions made over two decades of international engagement in Bamiyan, by an author (Mounir Bouchenaki) who has been devoted from the very start to the protection of that most precious heritage site. The next contribution offers a critical illustration of the various hypotheses of reconstruction of the Buddha statues advanced over the years by international experts, also highlighting the very distant heritage preservation outlooks that co-exist in the field (Hinz). Colombo and Panunzio focus on the cultural heritage found in the caves of Bamiyan's central cliff and lateral valleys, consisting mainly of the remains of Buddhist-era frescoes, providing an up-to-date assessment of the state of conservation two years after the interruption of cooperation interventions. The next two chapters deal with the sensitive issue of the protection of the cultural landscape, which was explicitly branded an object of protection in the 2003 nomination, yet remained on the margins of safeguarding actions for a long time. The first chapter (Toubekis) outlines the studies conducted between 2003 and 2013 on behalf of UNESCO by the University of Aachen/Germany, helmed by the late Michael Jansen, for the recognition and protection of Bamiyan's cultural landscape. The

---

[8] The conference casts the concept of sustainable development for the Afghan case into the term of modernization and opts for an "integration between cultural heritage preservation and modernization processes" (I.1). In general, the conference' resolution wishes to "show that the safeguarding of heritage properties and traditions is not a barrier to modernization, to the improvement of living conditions and to the economic growth of a country, but rather a driver for sustainable development" (Preamble).

second chapter (Loda, Di Benedetto) describes the specific measures proposed by urban planning tools—and in particular, by the *Strategic Master Plan* developed in 2018 by the University of Florence on behalf of the Italian Agency for Development Cooperation—for the protection of the Bamiyan cultural landscape, in an otherwise weak regulatory context regarding this aspect of heritage. In addition to providing details on the issues particular to Bamiyan's cultural landscape, the two chapters offer reflections on different approaches to the protection of the cultural landscape, both in practical and conceptual terms.

The second part of the volume is dedicated to contextualizing heritage protection within the broader framework of socio-cultural and local development dynamics, with examples that help develop guidelines for interventions in similar realities. Based on the Historic Urban Landscape approach (HUL), the first chapter illustrates the gradual emergence of integrated and potentially holistic approaches to heritage preservation in UNESCO practices (Francini, Rochozkina). This is followed by four chapters that illustrate the intertwining of preservation practices and other relevant local processes, and which use empirically collected primary data from within Bamiyan, in the period immediately prior to the Taliban's seizure of power. Hinz and Loda discuss ethnic tensions arising from the constraints introduced to safeguard cultural heritage, and the degree to which local populations have accepted these constraints; Loda, Amiri and Narayanan analyze the spillover of land titling policies in Zargaran (Bamiyan) and highlight their ability to facilitate participatory urban upgrading processes, even in contexts of prestigious cultural heritage; Loda, Di Benedetto and Potestà outline the pivotal elements for redevelopment policies aimed at urban districts subject to protection restrictions; Tartaglia and Ahmadzai examine the strategic role that accessibility to educational facilities plays, as part of the above-mentioned redevelopment policies. They then develop a specific method aimed at assessing the degree of educational sites' spatial accessibility. The second part of the volume concludes with a contribution focusing on the interaction between heritage preservation measures and urban development in Aleppo/Syria (Pucci). The third part of the book expands on how the conceptual tools described in the previous sections, inform operational approaches to the safeguarding of cultural heritage in fragile contexts by some of the main international actors in the field of international cooperation. Among these actors are UNESCO, ICCROM, Aliph, as well as the Italian Agency for Development Cooperation (AICS), which technically and financially supported both the Conference in Florence and the present publication.

The first three chapters by the Italian Agency for Development Cooperation describe the Agency's approach to culture as a tool of prosperity. That is to say, of improving livelihoods. The introductory chapter (Abenante-Strinati) describes an array of projects carried out by AICS in the Region focused on in this volume, to exemplify how, from a methodological and operational viewpoint, Italian actions are designed to generate economic activity and participation for and through culture. This approach places people, their own tools and objectives, at the center of cooperation practices, promoting ownership and inclusivity, and a form of cooperation that is seen as mutual learning and interaction between partners. The two following contributions re-employ this approach in Lebanon (Calia-Piermattei) and Jordan (Blasi), by focusing on the main initiatives carried out in the cultural sector and highlighting the impact of such initiatives on the tourism-driven economy of the two countries. The contributions continue by describing the main operational approaches of AICS' initiatives, such as the support to and upskilling of local professionals' capacities, the on-the-job training of unskilled workers, and the building of wide partnerships with both public institutions and civil society.

The contribution on the Shobak castle in Jordan (Nucciotti, Bala'awi, Sassu, Puppio, Candido) goes further into detail in describing the technical approach to site risk assessment, developed in one of the ongoing AICS funded projects in Lebanon. The authors explain how such an approach has been shared with local experts to allow priorities for long-term management to be established by the Department of Antiquities of Jordan.

The following paper (Khawam-Jigyasu) expands on ICCROM's approach to heritage education by focusing on the ICCROM Heritage Recovery Program in Mosul/Iraq, implemented in the framework of the UNESCO *Revive the Spirit of Mosul* initiative. The authors detail the stages of a capacity-building program on post-conflict heritage recovery that has proven valuable in improving livelihoods and empowering youth in the recovery of identity and of a sense of belonging through heritage. Operational approaches such as participatory need assessment, diversification of beneficiaries and intergenerational knowledge transfer have proven to be key success factors.

Hayashi's contribution describes the part that UNESCO has played in Afghanistan, from the 1950s to today. Beyond detailing UNESCO's actions, the contribution highlights the importance of memorialization processes that underpin the inscription of the Bamiyan Valley in the UNESCO World Heritage List. The contribution reminds the reader that reconstruction and restoration interventions will necessarily enter into a dialogue with memory. No matter how technically advanced, such interventions must consider the specific history of the site as well as the meaning and values communities give to the site, thus the connected living tradition.

The paper by Urtziverea describes the work of the International alliance for the protection of heritage in conflict area (Aliph) and its specific engagement in Iraq in heritage post-conflict reconstruction. The Alliance's actions, although carried out in emergency, are seen as efforts to foster collaboration and partnership with Iraqi institutions.

The last contribution of this section offers an up-to-date examination of international legislation in the field of cultural heritage protection (Greppi). The fourth part of the volume offers two contributions aimed at inferring methodological pointers for future interventions in the field of cultural heritage protection from the cases examined in the volume, both in the sphere of analysis and research (Loda) and in that of cooperation practices (Cabasino).

References

European Council (2005) Convention on the value of cultural heritage for society (Faro convention). https://www.coe.int/en/web/culture-and-heritage/faro-convention. Accessed 7 July 2007

ICCROM, ICOMOS, IUCN, UNESCO (2013) Managing cultural world heritage. World Heritage Resource Manual, UNESCO/ICCROM/ICOMOS/IUCN, Paris

ICOMOS (2021) Heritage and the sustainable development goals: policy guidance for heritage and development actors. ICOMOS, Paris. https://openarchive.icomos.org/id/eprint/2453/13/ICOMOS_SDGPG_2022%20-%20FINAL3.pdf. Accessed 25 Nov 2023

ICOMOS (2011) XVIIème assemblée générale. The Paris declaration on heritage as a driver of development. https://www.icomos.org/images/DOCUMENTS/Charters/GA2011_Declaration_de_Paris_EN_20120109.pdf. Accessed 19 Sept 2020

ICOMOS/Japan (2014) Nara + 20: on heritage practices, cultural values and the concept of authenticity. https://www.academia.edu/8972643/NARA_20_ON_HERITAGE_PRACTICES_CULTURAL_VALUES_AND_THE_CONCEPT_OF_AUTHENTICITY. Accessed 26 Jan 2016

ICOMOS/Japan (1994) The Nara document on authenticity. In: Nara conference on authenticity in relation to the world heritage convention, Nara, Japan. https://icomosjapan.org/static/homepage/charter/declaration1994.pdf. Accessed 3 May 2005

Islamic Republic of Afghanistan, World Bank Group, UNESCO (2016) Cultural heritage & development initiatives. A challenge or a contribution to sustainability?. Palazzo Barberini, Rome/ 25–27. https://www.unesco.org/culture/pdf/Symposium-Outcomes-Document.pdf. Accessed 8 Feb 2019

OECD (2022) The humanitarian-development-peace nexus interim progress review. OECD Publishing, Paris. https://doi.org/10.1787/2f620ca5-en

OECD (2022) States of fragility 2022. OECD Publishing, Paris. https://doi.org/10.1787/c7fedf5e-en

Smith LJ (2006) Uses of heritage. Routledge, London

UNESCO (2001) Universal declaration on cultural diversity, Paris 3. https://www.unesco.org/en/legal-affairs/unesco-universal-declaration-cultural-diversity. Accessed 8 Oct 2022

UNESCO (2022) The tracker culture & public policy|MONDIACULT Special issue n°5. https://www.unesco.org/en/articles/tracker-culture-public-policy-mondiacult-special-issue-ndeg5?hub=66775. Accessed 4 Nov 2023

UNESCO (2023) Operational guidelines for the implementation of the world heritage convention. https://whc.unesco.org/en/guidelines/. Accessed 29 Nov 2023

UNESCO (1972) Convention concerning the protection of the world cultural and natural heritage. https://whc.unesco.org/en/convention/. Accessed 24 July 1999

UNESCO (2003) Convention for the safeguarding of the intangible cultural heritage. https://ich.unesco.org/en/convention. Accessed 14 Aug 2023

UNESCO (2011) Recommendation on the historic urban landscape. https://whc.unesco.org/en/hul/#:~:text=About%20the%20Recommendation%20on%20the,in%20a%20changing%20global%20environment. Accessed 8 June 2022

UNESCO (2015) Policy for the integration of a sustainable development perspective into the processes of the world heritage convention. https://whc.unesco.org/en/compendium/55. Accessed 9 Nov 2019

Achievements and Challenges in the Protection of the World Heritage Site of Bamiyan

Heritage Preservation in Bamiyan: Achievements 2002–21

Mounir Bouchenaki

Abstract The contribution establishes the importance of the Florence conference on Cultural Heritage in Fragile Contexts. It firstly outlines the endeavors of various international organizations for the salvation of Bamiyan's giant Buddha statues as well as other assets of cultural heritage in the spring of 2001 and the role of the author in these initiatives. Then, it illustrates the bodies instituted by UNESCO for the management of the Bamiyan World Heritage site, which have been resumed by the above mentioned conference in Florence. Finally, it underscores the relevance of enduring capacity building for the safeguarding of cultural heritage assets, especially in fragile contexts.

Keywords Taliban · Iconoclasm · Bamiyan

1 The Terrible Fate of the Buddha Statues in Bamiyan Valley (Afghanistan)

"The giant Buddha statues of Bamiyan, the smaller Buddha probably dating back to the mid-sixth century AD, and the bigger one dating back to the early seventh century, are cut into the same cliff face with a distance of about 800 m between them. [...] The sculptures both show a standing Buddha. [...] Western literature distinguishes, according to position or size, between an Eastern or Small Buddha (38 m tall) and a Western or Big Buddha (55 m tall). Afghans identify the Eastern Buddha as *khinkbut* (grey or moon white Buddha) and the Western one as surkh-but (red Buddha)" (Blänsdorf and Petzet 2001, p. 18).

Some of the earliest records of the Bamiyan Buddhas can be found in Chinese travelogues. For example, in Da Tang Xiyu Ji's *Account of the country of Funyanna,* he describes the magnificent Buddha statues: "To the northeast of the royal capital is a mountain, at a secluded corner of which is a standing stone image of Buddha, one

M. Bouchenaki (✉)
UNESCO Expert, Paris, France
e-mail: mounir.bouchenaki@gmail.com

M. Loda and P. Abenante (eds.), *Cultural Heritage and Development in Fragile Contexts*, Research for Development, https://doi.org/10.1007/978-3-031-54816-1_2

17

hundred and forty or fifty *chis* (Chinese feet) high. Its golden hues are sparkling and its precious ornaments are glittering. To the east is a monastery, which was built by the last king. To the east of this monastery is a standing brazen figure of Sakyamuni Buddha, about one hundred *chi* high, different parts of the body being cast and joined together to get a complete form" (Nabi 2022, p. 61).

After the Islamic conquest, Arab authors were also familiar with the statues. "Nothing can be compared with these statues in the entire world," said Yakut al Hamawi about Bamiyan in his geographic dictionary, in the year 1218 (Blänsdorf and Petzet 2001, p. 19).

Thomas Hyde was the first European who mentioned the Bamiyan Buddhas in his writings in 1700, based on Arab literary sources. After nearly a century, Wilford and Elphinstone wrote about Bamiyan, also based on literary sources. It was William Moorcroft and George Trebeck who first went on an expedition to Afghanistan in 1824 and wrote about the Buddha statues in the niches. They were followed by the works of Alexander Burnes and Dr. Gerard.

"Do you prefer to be a smasher of idols or a seller of idols?" (Nabi 2022, p. 61) Mullah Omar asked Afghanistan's Muslim population on the 26th of February, 2001, when the Metropolitan Museum of Arts offered to pay for the Afghan artefacts.

Bamiyan's preservation and revival began with the 25th session of the Bureau of the *World Heritage Committee* on 25–30 June 2001 (UNESCO 2001a), in which the opening session included an extended discussion on the destruction of the Bamiyan statues. It quickly became a widely covered global issue which called for immediate action to be taken. Even though the quest to save Bamiyan by UNESCO started way before its demolition, and no amount of dialogue could convince the Taliban to do otherwise, the fight lost to the Taliban was certainly not the end.

2   The Value of the International Cooperation for Bamiyan Cultural Heritage

The Florence meeting[1] was a great opportunity to open the floor for a discussion with some of the main international sectorial organizations –UNESCO, ICOMOS, Aga Khan Trust for Culture, ALIPH—and several Italian universities, as well as with the Agenzia Italiana per la Cooperazione allo Sviluppo (AICS) offices, operating in the Middle East and the Indian subcontinent.

A whole session was entirely dedicated to the World Heritage site of Bamiyan, in Afghanistan, at the center of a redevelopment project that sees the *Dipartimento di Storia, Archeologia, Geografia, Arte e Spettacolo* (SAGAS) of Florence University as a partner of AICS—Islamabad, following the closure of AICS Kabul in 2021. This session examined the main cultural heritage protection efforts carried out in the Bamiyan area, with the aim of discussing the results achieved prior to the Taliban

---

[1] *Cultural Heritage in Fragile Contexts. Development Cooperation in Afghanistan and Neighboring Countries*, Florence University, November 10–11, 2022.

coming back into power, in an attempt to highlight the most critical issues and sectors in need of further interventions.

This was among the objectives of this conference, planned by Professor Mirella Loda, who told *The Art Newspaper*: "I deeply hope we will be able to set up a group of international experts that could orient local authorities so as, at least, to prevent destructive and irreversible interventions" (Geranpayeh 2022). She was speaking as project coordinator for Bamiyan's *Strategic Master Plan*.

Let us now start with the UNESCO approach to the Bamiyan Valley.

"UNESCO defines the *Cultural Landscape and Archaeological Remains of the Bamiyan Valley* as representing the artistic and religious developments which characterized ancient Bakhtria from the 1st to the 13th century, integrating various cultural influences into the Gandhara school of Buddhist art. The area contains numerous Buddhist monastic ensembles and sanctuaries, as well as fortified edifices from the Islamic period. The site is also testimony to the tragic destruction by the Taliban of the two standing Buddha statues, which shook the world in March 2001" (UNESCO World Heritage Centre 2003).

As mentioned in the book published by the UNESCO Office in Kabul in 2005 (UNESCO 2005), much discussion has taken place in Afghanistan and globally, regarding the future of this site, revolving around the question of whether the two statues of Buddha should be reconstructed. Already, very early after the fall of the Taliban in 2002, a UNESCO expert fact-finding mission traveled to Bamiyan in order to examine the situation. This mission consisted of Prof. Michael Jansen of RWTH Aachen University, Prof. Michael Petzet, President of the International Council on Monuments and Sites (ICOMOS), Prof. Kosaku Maeda of the National Research Institute of Cultural Properties Tokyo (NRICPT), among other international experts. The technical aspects of the mission revealed that under the rubble, many fragments from the destroyed figures could be found with original surface features. While the detonations destroyed the Western Buddha almost entirely, portions of the Eastern Buddha survived the explosion, as reported by ICOMOS expert Santana-Quintero in 2002.

A plan for the preservation of these fragments and the long-term conservation of the remains was first presented by Professor Michael Petzet, immediately after the international expert meeting I had the responsibility of organizing in Kabul in May 2002.

The 107 participants at the First International Seminar on the Rehabilitation of Afghanistan's Cultural Heritage, organized by UNESCO in May 2002, as well as the *International Coordination Committee for the Safeguarding of Afghanistan's Cultural Heritage* (ICC), clearly recognized that the main emergency priority is to stabilize the cliff face with its niches and caves. Noting that the decision to engage in the reconstruction of the Buddha statues is a matter to be settled by the government and the people of Afghanistan, it was agreed that the reconstruction of the Buddha Statues would not be a priority for as long as humanitarian aid for the Afghan people was urgently needed.

Furthermore, it was emphasized that the authenticity, integrity, and historical importance of this great site needed to be memorialized appropriately, and that because of this, the reconstruction of the statues would require further careful consideration. The preservation of the Bamiyan Site was one of the most important UNESCO projects in Afghanistan, for which more than $ 1.8 million have been generously donated by the Government of Japan.

The *Cultural Landscape and Archaeological Remains of the Bamiyan Valley* was inscribed on the "List of World Heritage in Danger" and the World Heritage List at the 27th session of the World Heritage Committee in 2003. The property is in a fragile state of conservation, having suffered from neglect, military action, and dynamite explosions. In 2003, the major dangers included the risk of imminent collapse of the Buddha niches with the remaining fragments of the statues, further deterioration of still-existing mural paintings in the caves, looting, and illicit excavation (UNESCO 2005).

Architect Paul Bucherer, who is very familiar with the cultural heritage of Afghanistan, also took part in the first missions undertaken after the destruction of the giant Buddhas. I had the pleasure to travel with him to Kabul. I was also present with him in New York in April 2003 for a panel discussion organized by the *Asia Society* entitled, "Beyond Bamiyan: Will the World be Ready Next Time?" The panel took place in order to address current debates on the issue of cultural heritage in Afghanistan, and the steps taken by the international community to protect it. This program was made possible by the generous support of the Hazen Polsky Foundation (Asia Society 2003).

I was among the panelists, alongside Mrs. Bonnie Burnham, President of the World Monuments Fund, Mrs. Barbara Crossette, a reporter for The New York Times, Mr. James Cuno, Director of the Harvard University Art Museums, Mr. Philippe de Montebello, Director of the Metropolitan Museum of Art, Mr. Derek Gillman, President of the Pennsylvania Academy of the Fine Arts, and Mr. Satoshi Yamato of the Agency for Cultural Affairs in Japan.

Architect Paul Bucherer is the founder and director of the Afghanistan Institute and Museum (*Bibliotheca Afghanica*) in Switzerland. The *Bibliotheca Afghanica Museum-In-Exile,* which he created in Bubendorf near Basel, was serving as a temporary home for artifacts that were loaned to Mr. Bucherer by Afghans and others outside the country, for safekeeping. As an activist, Paul Bucherer has spent more than three decades documenting the country's cultural heritage and has supplied the *Afghanistan Museum in Exile* and the *Bibliotheca Afghanica* with numerous artifacts. He was providing UNESCO with critical information on the state of sites before, during, and after the civil unrest, while institutions that once worked in Afghanistan were beginning to return to their sites, in order to resume their study and conservation activities.

Between 2000 and 2006, my colleagues and I were at Headquarters and at UNESCO's office in Kabul, in charge of following up the tragic destruction of the Buddhas of Bamiyan.[2] In this position, I had regular contact with Mr. Bucherer, but also with Professor Ikuo Hirayama, a celebrated contemporary artist, working in

---

[2] In Paris, Mr. Christian Manhart, Program Specialist and in UNESCO Kabul office, Mr. Jim Williams and Mr. Brendan Cassar.

the genre of Nihonga, and UNESCO Goodwill Ambassador and special adviser for cultural heritage, specially appointed director of the Tokyo National Museum.

Both facilitated the return of many works of art to Afghanistan, several of which they had gathered in various countries, with the protection in Switzerland of the Swiss Parliament, and in Japan thanks to the highly respected position of Professor Ikuo Hirayama, who advocated Red Cross activities for cultural heritage.

UNESCO's experiences through events in Cambodia, Southeast Europe, Afghanistan, Iraq, the Middle East, Timor-Leste, and elsewhere, provide hope that a program can be put in place for the preservation of cultural heritage, aiming at reconstruction on one hand, and dialogue and reconciliation on the other.

UNESCO has always held that archaeological sites, in addition to old manuscripts, give unique insight into civilizations. They are frequently associated with ideas or beliefs that have marked the history of humanity from time immemorial.

To respond to the numerous queries that UNESCO received following the destruction of the Bamiyan Buddhas, and to put an end to false interpretations of Islamic law concerning cultural heritage, while also preventing such iconoclastic acts from occurring in the future, a conference of specialists in Islamic law was organized in Doha (Qatar), from the 29th to the 31st of December, 2001, on the occasion of the Ministers of Culture of the Islamic World regular meeting. We started the preparations for this Conference on the 15th of March, 2001, immediately after the destruction of the Bamiyan Buddha statues. Contact was made with the most renowned specialists (Ulamâ) in Islamic law (Sharia) from different religious schools (Sunna and Shia). Specialists from Morocco, in the western Islamic World, and up to Kazakhstan, in its Eastern part, were invited.

The agenda of the conference went beyond safeguarding the Buddha statues in Bamiyan and the sustainable protection of Afghan cultural heritage. Indeed the Doha Conference of "Ulamâ", entitled *Islam and Cultural Heritage*, was chaired by His Highness Sheikh Hamad bin Khalifa Al Thani, Emir of the State of Qatar, and put together by the Organization of the Islamic Conference (OIC), the Islamic Educational, Scientific and Cultural Organization (ISESCO), the Arab League Educational, Cultural and Scientific Organization (ALECSO), and UNESCO (UNESCO 2001b).

It was inaugurated by the directors of three international and regional organizations: Mr. Koïchiro Matsuura (UNESCO), Mr. Abdulaziz Othman Altwaijri (ISESCO), and Mr. Mongi Bousnina (ALECSO). It should be noted that the conference was attended by 27 professors and experts in Islamic Law, from 25 different countries. A delegation from Afghanistan also participated, led by Professor Sibghat-ullah Mujaddidi, former President of Afghanistan. After 2 days of intense discussion, this meeting resulted in the "Declaration of Doha", later widely circulated throughout the Islamic world.

The "Ulamâ", in attendance discussed the various aspects of the Symposium with a special focus on the recent destruction of the Buddhas in the Bamiyan Valley. They stressed that the tolerant nature of Islam demands respect for human heritage in general, whatever its origins, forms, or manifestations. In their deliberations, they highlighted the fact that Muslims have preserved human heritage in all its diversity, making sure not to harm it in any way. This is evidenced by the fact that the

Islamic world boasts the greater part of human heritage, most of which goes back to pre-Islamic periods, notably in the Middle East and North Africa. Had it not been preserved by Muslims, most of those heritage sites would have been lost. The Ulamâ noted that the situation has remained so over the entirety of Islam's 14th-century history.

The Ulamâ participating in the DOHA Symposium affirmed that the "position of Islam with regard to the preservation of the human cultural heritage derives from its appreciation of innate human values and from respect for people's beliefs". They explained that "the position of Islam regarding the preservation of the cultural heritage is a firm position of principle which expresses the very essence of the Islamic religion" (Bouchenaki 2020, p. 26).

3   International Financial and Technical Cooperation for Bamiyan

Following the UNESCO-sponsored meeting in May 2002, numerous other conservation initiatives have been planned. In the historic center of Kabul, the Aga Khan Trust for Culture (AKTC) has embarked on a major campaign to restore a number of eighteenth and nineteenth-century *serays* and residences, as well as the famed seventeenth-century Babur Gardens in the northwest part of the city. AKTC, with support from the World Monuments Fund (WMF), is resuming restoration efforts in Herat, a fifteenth-century Timurid city, which is included on WMF's 1998 list of the 100 Most Endangered Sites.

In Bamiyan, Michel Petzet, President of ICOMOS and his colleagues have been documenting what is left of the 1,500-year-old giant Buddhas and the fragmentary murals that once graced the hundreds of caves in the valley.

Elsewhere in the country, Mrs. Nancy Hatch Dupree and her team at the Society for the Preservation of Afghanistan's Cultural Heritage (SPACH), have been undertaking detailed assessments of heritage conditions, as well as reconstructing museum inventories.

Collectively, these efforts, along with programs undertaken by government agencies, are working to rebuild one of the world's great cultural crossroads.

In 2002, the *Bamiyan Expert Working Group* (BEWG) was established, and the Government of Afghanistan entrusted UNESCO with coordinating all cultural projects across Bamiyan. The BEWG also advises the Government of Afghanistan in implementing the decisions adopted by the World Heritage Committee for the World Heritage property of Bamiyan in the conservation and management areas.

The conservation activities were started in 2003 with the UNESCO Campaign that Mr. Koichiro Matsuura, Director General of UNESCO, launched for the Preservation of the Bamiyan Site. Part of the funding was provided by Italian, Japanese, and German authorities, in addition to the financial assistance which the UNESCO World Heritage Centre made available, upon the site's inscription in 2003.

With regard to the official installation by UNESCO of an *International Coordination Committee for the Safeguarding of Afghanistan's Cultural Heritage* (ICC) which was completed in 2002, the objectives were stated as follows: "Responding to the urgent need to enhance and facilitate the coordination of all international activities, and in accordance with the Afghan authorities, UNESCO has established an International Coordination Committee (ICC). Its statutes were approved by the 165th session of the organization's Executive Board in October 2002" (ICC 2004).

The ICC, which consists of Afghan experts and leading international specialists belonging to those donor countries and organizations who are doing most to provide funding or scientific assistance towards safeguarding Afghanistan's cultural heritage, meets on a regular basis to review ongoing and future efforts to rehabilitate that heritage. In June 2003, the Committee's First Plenary Session was organized at the UNESCO headquarters in Paris.

It was chaired by Makhdoom Raheen in the presence of Prince Mirwais, 7 representatives of the Afghan *Ministry of Information and Culture*, and more than 60 international experts participating as members of the Committee or as observers. The meeting resulted in a set of specific recommendations for the efficient coordination of efforts to safeguard Afghanistan's cultural heritage, to the highest international conservation standards.

"These recommendations concern key areas such as the development of a long-term strategy, capacity building, implementation of the World Heritage Convention and the Convention on the Means of Prohibiting and Preventing the Illicit Import, Export, and Transfer of Ownership of Cultural Property, national inventories, and documentation, as well as rehabilitation of the National Museum in Kabul and safeguarding of the sites of Jam, Herat, and Bamiyan. Several donors pledged additional funding for cultural projects in Afghanistan following the meeting" (Manhart 2004, p. 403).

The consolidation of the extremely fragile cliffs and niches, the preservation of the mural paintings in the Buddhist caves, and the preparation of an integrated master plan, were prioritized. In order to prevent the collapse of the cliffs and niches, large scaffolding was supplied free of charge by the German Messerschmidt Foundation and transported by the German army to Afghanistan in August 2003. With the help of this scaffolding and other imported specialized equipment, the internationally renowned Italian firm RODIO successfully completed the first phase of the emergency consolidation of the cliffs and niches.

The *Bamiyan Expert Working Group* was also established in 2002 and has as its main goal the coordination of all activities carried out in Bamiyan under the various UNESCO projects, as well as any bilateral activities funded by international donors. It also advises the Government of Afghanistan on the implementation of decisions adopted by the World Heritage Committee for the World Heritage property of Bamiyan in the areas of conservation and management.

In July, September, and October 2003 several missions, consisting of specialists from the National Research Institute for Cultural Properties (Japan), were sent to

Bamiyan to safeguard the mural paintings and to draw up a master plan for the long-term preservation and management of the site. A Japanese firm was commissioned to prepare a topographical map of the valley and a 3D model of the cliffs and niches.

"A Japanese survey team by PASCO Inc. set up the coordinates for a site reference system and produced a detailed topographic map of the central Bamiyan valley based on ground-trusted analysis of Quickbird satellite data and aerial images provided by the Afghan Geodesy and Cartography Head Office (AGCHO). The PASCO team also realized a 3D laser scan of the entire cliff, the niches of the Giant Buddhas and several caves in order to document the condition after the detonation and to prepare site plan material for the next stages of the program" (Toubekis et al. 2009).

In addition, UNESCO helped the Afghan government to create a site museum, planned to be set up in a traditional house near the site. To this end, the Swiss government approved a UNESCO Funds-in-Trust project for the restoration of a traditional mud-brick house in the village of old Bamiyan, with a budget of US$ 250,000 for the study of traditional houses to be carried out, so that appropriate restoration methods could then be recommended.

In January 2003, the Greek government started the restoration of the Kabul Museum building, as part of a commitment made during the Kabul Seminar in May 2002, consisting of a donation of approximately US$ 750,000. UNESCO provided the Greek specialists with drawings of and plans for the Kabul Museum, produced by the organization's consultant, Architect Andrea Bruno. The US government also contributed US$ 100,000 to this project.

Further to this, the British International Security Assistance Force (ISAF) installed a new restoration laboratory composed of two rooms, one wet-room and one dry-room, both of which were funded by the British Museum. In addition, the French *Centre d'Études et de Recherches Documentaires sur l'Afghanistan* (CEREDAF) donated conservation equipment, while the newly re-established French *Délégation Archéologique Française en Afghanistan* (DAFA), alongside the Guimet Museum in Paris, carried out training courses for the museum's curators initiated by the Italian firm IsIAO in 2002 (Manhart 2004, p. 409).

The 12th Bamiyan Expert Working Group Meeting was held at the Technical University of Munich, Germany (1st–3rd of December 2016), and provided important recommendations for future safeguarding actions at the Bamiyan World Heritage site. The Governor of the province of Bamiyan, the Deputy Minister of Culture of Afghanistan, accompanied by high-ranking officials from the Government of Afghanistan, attended the meeting, along with representatives of donor countries, such as Germany, Italy, Japan, and the Republic of Korea, along with 20 international experts from Italy, Germany, and Japan.

The 14th Bamiyan Expert Working Group Meeting, held in Tokyo, Japan, in 2017, was especially important. The Tokyo meeting initially assessed the progress made with regard to the conservation of the property and then laid out priority activities for the immediate future in the form of recommendations. These recommendations (UNESCO 2017) included the urgent conservation of the Western Buddha niche, where conservation works had resumed in the second semester of 2016, as well as the need for a revision of the Bamiyan *Cultural Master Plan*, originally developed in 2004, in view of ever-increasing development pressures.

The meeting also served as a platform for preliminary discussions between the Government of Afghanistan, international experts, and donor countries on the feasibility of reconstructing at least one of the Buddha statues, which has been officially requested by the Government of Afghanistan on behalf of the people of Afghanistan (Bouchenaki 2020, p. 28).

The issue of producing a *Cultural Master Plan* for the Bamiyan Valley was a concern for the late Professor Michael Jansen.

On the 8th of December 2020, he published an article entitled "The Cultural Master Plan of Bamiyan: The Sustainability Dilemma of Protection and Progress" with Dr. Giorgos Tubekis (Jansen and Toubekis 2020). The authors underline the fact that "beyond the Buddha Cliff, the World Heritage property of Bamiyan consists of several archaeological areas embedded into an extraordinary cultural landscape not adequately defined at the time of the nomination. Therefore, the *Cultural Master Plan* was envisioned as guidance for the development of a rural environment under cultural preservation objectives. The plan introduces a zoning scheme defining land use regulations for the protection of cultural areas, and proposing designated areas for urban development. Lack of adequate legal protection, too-rigid enforcement of land use restrictions on the local level, and the aspirations of the people for a rapid change of their living conditions resulted in increasing uncertainties on the validity of the plan."

It is further argued that a monitoring guiding committee composed of international and national experts, as well as local stakeholders, would be helpful in counterbalancing uncoordinated international aid assistance and inefficient governmental supervision, which had led to overemphasizing urbanization approaches in development strategies, something that was conflicting with the rural character of the valley. The authors propose a reconsideration of urbanization within the Bamiyan Valley and a reconciliation of the objectives of urban and rural development, inspired by a sustainable development plan as put forward by the Sustainable Development Goals (SDGs) of the UN 2030 Agenda for Sustainable Development.

As we can see, during the last 20 years, international cooperation has been very active in the Bamiyan valley, while many countries have contributed to the protection of the area's sites in various ways, and not only the two niches where the Buddhas were destroyed.

The Italian and Afghan governments have cooperated with UNESCO on heritage protection for more than five decades, thanks largely to the work of Italian organizations and universities. The key results of this cooperation were presented at the November 2022 seminar, held at the University of Florence.

4 Training and Capacity Building in Afghanistan Post-Conflict Period

Apart from the work done in the aftermath of the Buddhas' destruction, there is another strategic field which was addressed by the participants of the meeting in Florence, namely the training of young Afghan professionals in the protection and presentation of their country's rich cultural heritage. Apart from the University of Florence, the University of Arizona also became involved: "Archaeologists estimate there are more than 5000 archaeological sites in Afghanistan, but not all have been identified. And after three decades of war and neglect, Afghan sites and artifacts are in serious need of trained staff, proper site management, security, funding, national and international commitment, and development of advocacy and awareness programs" (College and of Architecture, Planning and Landscape Architecture 2014).

In 2022, Richard Mulholland published a report on the HUNAR (Heritage Unveiled: National Art Restoration) Program and highlighted the importance of training for Afghans (Mulholland 2002): "One of the HUNAR program's key objectives in 2020–2021 was to address the lack of skills and knowledge in heritage management and conservation in Afghanistan. This included the design and implementation of a simple collections management database at the Afghan National Gallery, the translation of key conservation sources on paper and easel paintings conservation into Dari and Pashto, and an intensive training course for participants from the National Gallery, National Museum, National Archives, Kabul University, the Art Institute, and four provincial galleries.

The training took the form of a short, ten-day course on the theory and practice of collections management, conservation, artists' materials and techniques, and preventive conservation, to be put into practice in Kabul in 2020. However, the advent of COVID-19 combined with a worsening security situation in Kabul meant that travel to the region became impossible, and the training was moved online. The course was provided via recorded lectures with live translation, followed by a live online 'Q&A' session with all participants.

A small group of high-scoring trained participants were then selected to travel to regional galleries at Kandahar, Herat, Balkh, and Nangarhar to train local staff in condition reporting and basic collections management. Social and gender inclusion was promoted strongly throughout the project. As with most sectors in Afghanistan, there are structural inequalities that make accessing decision-making and leadership roles in the heritage sector difficult. In most provinces, there are no female employees in the sector. HUNAR was the first project of its kind to carry out a formal gender and social inclusion assessment for the heritage sector in the region."

In Afghanistan, an obvious international sustainable development goal is to promote the empowerment of young women (Wimpelmann 2017), and the inclusion of heritage in the international development agenda presented a real opportunity to re-evaluate the role of women in contemporary Afghanistan, especially since women are over-represented in higher education art and design courses, which is the traditional entry route into heritage roles (Hashimi 2021). The Afghan Ministry of

Culture employs 2023 people, 14% of whom are female. However, the vast majority of women work in the capital. The ratio is almost non-existent in the provinces, where there is also in general very low education attainment for women.

It is important to state that there is no common narrative for cultural heritage and the people that it represents. Richard Mulholland's limited study has demonstrated that although international conservation interventions in post-conflict zones can have a significant impact, it can be challenging to achieve a sustainable impact in the cases where cost and security issues generally mean that training takes place over an intensive but brief period of time (Mulholland 2002).

In concluding this brief introduction on the Florentine Conference on *Cultural Heritage in Fragile Contexts*, we should refer to the UNESCO declaration commemorating the 20th anniversary of the destruction of the Buddhas of Bamiyan on the 11th of March 2021.

"Although the destruction of heritage and the plundering of artefacts have taken place since antiquity, the destruction of the two Buddhas of Bamiyan marks an important turning point for the international community. A deliberate act of destruction, motivated by an extremist ideology that aimed to destroy culture, identity, and history, the loss of the Buddhas revealed how the destruction of heritage could be used as a weapon against local populations. It highlighted the close links between heritage safeguarding and the well-being of people and communities. It reminded us that defending cultural diversity is not a luxury, but rather a necessity in building more peaceful societies. Since the destruction of the Buddhas of Bamiyan, the Afghan authorities and the international community, including UNESCO, have worked tirelessly to safeguard the rich cultural and natural heritage of Afghanistan, which testifies to millennia of exchanges between different cultures and peoples" (UNESCO 2021).

References

Arizona College of Architecture, Planning and Landscape Architecture (2014) Professional education for Afghan cultural heritage faculty. Final report, https://capla.arizona.edu/sites/def ault/files/Assistance%20with%20Professional%20Training%20for%20Afghan%20Cultural% 20Heritge%20Officials%20Pt%202.pdf. Last Accessed 07 Dec 2018

Asia Society (2003) How can Afghanistan's cultural heritage be preserved? Interview with Paul Bucherer, https://asiasociety.org/how-can-afghanistans-cultural-heritage-be-preserved. Last Accessed 06 Nov 2004

Blänsdorf C, Petzet M (2001) Description, history and state of conservation before the destruction in 2001. In: Petzet M (ed) ICOMOS XIX, The Giant Buddhas of Bamiyan. Safeguarding the remains, pp 17–29. Bäßler Verlag, Berlin

Bouchenaki M (2020) Safeguarding the Buddha statues in Bamiyan and the sustainable protection of Afghan cultural heritage. In: Nagaoka M (ed) The future of the Bamiyan Buddha statues. Heritage reconstruction in theory and practice, pp 19–30. UNESCO-Springer, Paris

Geranpayeh S (2022) Should the world resume co-operating with the Taliban on protecting Afghanistan's heritage? Conservation projects that have been paused due to sanctions on the new government may restart after UNESCO intervention, The art newspaper, interview with

Professor Mirella Loda, dated 25th of November 2022, https://www.theartnewspaper.com/2022/11/25/should-the-world-resume-co-operating-with-the-taliban-on-protecting-afghanistans-heritage. Last Accessed 25 Jan 2023

Hashimi SS (2021) Factors impacting work-life balance of female employees in private higher education institutions in Afghanistan: an exploration. Kardan J Econ Manag Sci 4(3):84–99. https://doi.org/10.31841/KJEMS.2021.101

International Coordination Committee for the Safeguarding of Afghanistan's Cultural Heritage (ICC) (2004), https://whc.unesco.org/en/activities/245/. Last Accessed 28 Nov 2004

Jansen M, Toubekis G (2020) The cultural master plan of Bamiyan: the sustainability dilemma of protection and progress. In: Nagaoka M (ed) The future of the Bamiyan Buddha statues. Heritage reconstruction in theory and practice, pp 71–98. UNESCO-Springer, Paris

Manhart C (2004) UNESCO's mandate and recent activities for the rehabilitation of Afghanistan's cultural heritage. Int Rev Red Cross 86(854):401–414

Mulholland R (2002) Culture, education and conflict: the relevance of critical conservation pedagogies for post-conflict Afghanistan. Stud Conserv 67(2):283–297

Nabi A (2022) Crossroad of culture. A contested space of culture and cultural terrorism–Bamiyan, Afghanistan. Int J Arts Soc Sci 5(11):60–68

Toubekis G, Mayer I, Doring-Williams M, Maeda K, Yamauchichi K, Taniguchichi Y, Morimoto S, Petzet M, Jarke M, Jansen M (2009) Preservation and management of the UNESCO world heritage site of Bamiyan: laser scan documentation and vitual reconstruction of the destroyed Buddha figures and the archeological remains, https://publications.rwth-aachen.de/record/113884/files/113884.pdf?subformat=pdfa. Last Accessed Apr 2018

UNESCO (2001a) Twenty-fifth session of the Bureau of the World Heritage Committee to meet 25–30 June at UNESCO headquarters, Paris. UNESCO World Heritage Centre, https://whc.unesco.org/en/news/168/. Last Accessed 15 May 2022

UNESCO (2001b) Proceedings of the Doha conference of 'Ulama on Islam and cultural heritage, Doha, Qatar, 30–31 December 2001, https://unesdoc.unesco.org/ark:/48223/pf0000140834. Last Accessed 26 Sept 2004

UNESCO (2005) Islamic Republic of Afghanistan, Ministry of Information and Culture, Bamiyan, Preserve our Cultural Heritage, Kabul

UNESCO (2021) Commemorating 20 years since the destruction of two Buddhas of Bamiyan, Afghanistan, https://whc.unesco.org/en/news/2253. Last Accessed 15 May 2021

UNESCO World Heritage Centre (2003) Cultural landscape and archaeological remains of the Bamiyan Valley, nomination dossier 208kev

Wimpelmann T (2017) The pitfalls of protection. Gender, violence, and power in Afghanistan. University of California Press, Oakland

The Fate of Bamiyan's Giant Buddha Statues

Manfred Hinz

Abstract This contribution outlines the debate around a possible and (from the Afghan side until 2021) sought after reconstruction of Bamiyan's giant Buddha statues, which started after their destruction in 2001, and lasted up until the UNESCO conference on the subject in 2017. This debate gains significance against a background of ongoing discussion concerning the legitimacy of heritage reconstruction in general, which has greatly altered the key concepts of authenticity and integrity. This general discussion is, naturally, still open. But consideration of the Bamiyan case has been rendered obsolete by the political events of August 2021. This contribution advances the opinion that a physical reconstruction of the statues would be redundant, since their destruction has established them firmly and universally in the collective imagery.

Keywords Iconoclasm · Heritage reconstruction · Anastylosis

1 Premise

The debate surrounding the possible reconstitution, revivification, reassembly, anastylosis and even complete reconstruction of Bamiyan's two giant Buddha statues (38 and 55 m) started right after their destruction in March 2001. This debate involved technical and theoretical issues, given that the international heritage preservation community, UNESCO, ICOMOS, ICCROM etc., had already committed itself to strict technical, theoretical and financial limitations. The milestone achievements in this effort at stringency are the *Venice Charter* of 1964, the *World Heritage Convention* of 1972, and the *Nara Document on Authenticity* of 1994, whose warranties had

M. Hinz (✉)
Philosophische Fakultät, Passau University, Passau, Germany
e-mail: manfred.hinz@uni-passau.de

Department of History, Archaeology, Geography, Fine and Performing Arts, Laboratory for Social Geography (LaGeS), University of Florence, Via San Gallo 10, 50129 Firenze, Italy

© The Author(s) 2024
M. Loda and P. Abenante (eds.), *Cultural Heritage and Development in Fragile Contexts*, Research for Development, https://doi.org/10.1007/978-3-031-54816-1_3

29

to be reconsidered or worked around alongside strictly technical questions regarding the remnants of the two Buddha statues and their now empty niches.

Despite a remarkable commitment by the international heritage preservation community, this debate is still open on all sides involved. The state partner, the Islamic Republic has continuously demanded (until the events of August 2021) some sort of reconstruction work on one or both the Buddha statues. The international consultant bodies have withheld making a decision, have had reservations and put a number of technical proposals on the table. The theoretical debate concerning the criteria of reconstruction which safeguards the authenticity and integrity of a site as demanded by UNESCO has grown truly global since 2001, the oxymoron of "authentic reconstruction" has enticed the attention of academics around the world. The problem of funding, obviously, remains fundamental.

In this contribution, we will recall the debate surrounding the possible reconstruction of Bamiyan's giant Buddha statues, contextualizing the proposals brought up within the framework of the broader contemporary discussion on the recovery of lost cultural heritage. We will start by briefly reconsidering the events and political decisions that led to the destruction of the Buddha statues (as well as many other heritage assets in Afghanistan), so as to understand their significance, both intended and real. Only an assessment such as this allows for an estimation of how the local population and their political representatives might react to any intervention by the international heritage preservation institutions. The second section reviews the two-decade-long debate on the future management of the remnants of the two Buddha statues, its main actors being UNESCO and its advisory bodies on the one hand, and the Afghan state party on the other. The third section discusses the technical proposals exhibited at the UNESCO conference of 27–29 September 2017 in Tokyo. In the fourth part of the article, we offer some concluding remarks.

2   The Destruction

In July 1999, Mullah Omar, the Supreme Leader of the first Islamic Emirate of Afghanistan issued a decree whose first point reads: "All historical cultural heritages are regarded as an integral part of the heritage of Afghanistan and therefore belong to Afghanistan, but naturally also to the international community. Any excavation or trading in cultural heritage objects is strongly forbidden and will be punished in accordance with the law" (Falser 2011, p. 159).[1] In 1979, Afghanistan had in fact adhered to the UNESCO *Convention Concerning the Protection of the World Cultural and Natural Heritage* (1972) and although the country lacked a World Heritage nomination in those years, the Taliban explicitly recognized the authority of the international bodies over its cultural heritage.

This decree, as well as early Taliban policy in general, were effective. Indeed, the relevant volume of the *Handbook for Oriental Studies* asserts: "Contrary to what

---

[1] For the context of this decree see Cassar and García (2006, p. 34).

many people think, the illegal excavations and the illicit trade in antiquities did decline sharply during the time of the Taliban rule" (Van Krieken-Peters 2006, p. 206). Following this decree, even the National Museum of Afghanistan in Kabul, formerly devastated and heavily looted over the years of the Mujahidin civil war, prior to the Taliban's rise to power, could briefly reopen in August 2000 (Grissmann 2006).

Bamiyan is explicitly mentioned at length in point six of Mullah Omar's decree quoted above: "The famous Buddhist statues at Bamiyan were made before the event of Islam in Afghanistan, and are amongst the largest of their kind in Afghanistan and in the world. In Afghanistan, there are no Buddhists who worship the statues. Since Islam came to Afghanistan, until the present period, the statues have not been damaged. The government regards the statues with serious respect and considers the position of their protection today to be the same as always. The government further considers the Bamiyan statues as an example of a potential major source of income for Afghanistan from international visitors. International Buddhist communities recently issued a warning that in the case, the Bamiyan statues are damaged, then mosques will be damaged in their regions. The Muslims of the world are paying attention to this declaration. The Taliban government states that Bamiyan shall not be destroyed but protected" (Falser 2011, p. 159). Already in 1999 there evidently existed tensions around the Bamiyan Buddha statues.

During the 21st World Heritage Committee meeting in Naples, December 1–6, 1997 (WHC-97/CONF.208/17), shortly after the Taliban took control of Bamiyan, at the request of the Japanese delegate, the WHC expressed its concern for "news reports on threats to the cultural and natural heritage of Afghanistan, particularly the Buddhist statues in Bamiyan", and invited "the authorities in Afghanistan to take appropriate measures in order to safeguard the cultural and natural heritage of the country" (UNESCO 1998, VII.58, p. 34). UNESCO's plea was directed to a government recognized only by Pakistan, Saudi Arabia and the Arabic Emirates, and had no seat in the UN (which was occupied by a representative of the previous Rabbani government). However, whether as a reaction to the plea or not, the Islamic Emirate in its search for larger international recognition was keen to assure the world community that it would protect its national cultural heritage, paying particular attention to the Bamiyan Buddhas, as documented in Mullah Omar's 1999 decree.

Hopes for international recognition came to nothing. The country's economic crisis, exacerbated by a terrible drought (1999–2001), plagued the country, causing widespread deaths. The political climate also changed rapidly.[2] While the Supreme Leader's decrees of 1999 passed largely unnoticed, his decree broadcasted on Radio Shari'at, 1 Hut 1421 (26 February 2001) and two days later in Afghan newspapers, announcing the destruction of Bamiyan's Buddha statues and other pre-Islamic assets in the country, caused an immediate worldwide outcry: "Based on consulting religious leaders of the Islamic Emirate of Afghanistan, the religious Ulema's and the Islamic Emirate of Afghanistan Supreme Court's judgment, all the statues in the different parts of the country must be broken because these statues have remained

---

[2] For the political history of the first Islamic Emirate of Afghanistan, see Van Linschoten and Kuehn (2012).

as shrines for infidels and they are worshipping these statues and still the statues are being respected [by them] and probably they will be changed to shrines again, while God Almighty is the real shrine and all the false shrines must be smashed. Therefore, the authorities of the Islamic Emirate of Afghanistan have given the duty to the Ministries for the Promotion of Virtue and the Prevention of Vice as well as Information and Culture to destroy all the statues in order to implement the judgments of the Ulema and the Supreme Court. All the statues must be annihilated so that no one worships or respects them in the future".[3]

This new decision was swiftly implemented. By the 12th of February 2001, the BBC had already reported that Taliban militias had started to destroy antique statues in the National Museum in Kabul. The demolition of the Bamiyan Buddhas turned out to be a more complicated affair than expected, taking several weeks. On the 2nd of March, the Afghan Islamic Press reported that explosives had been collected around the country and transported to Bamiyan. On the 10th of March, the same news agency reported that detonations had resumed after having been interrupted for the Eid-al Adha holiday. On the 11th of March, Al Jazeera broadcast the blasting of the Buddha statues to appalled audiences worldwide. The Taliban government was evidently eager to boast about its achievements to the international public. On the 22nd of March, a group of journalists was invited to a then looted National Museum in Kabul and on the 26th of March, another group was flown to Bamiyan to report on the empty Buddha niches.[4] A Taliban envoy was sent to the United States and interviewed on American TV on the 18th of March. According to the interview, published by the *New York Times* a day later, "The Islamic government made its decision in a rage after a foreign delegation offered money to preserve the ancient works while a million Afghans faced starvation. (…) The destruction … was prompted when a visiting delegation of mostly European envoys and a representative of the UNESCO offered money to protect the giant standing Buddhas at Bamiyan. (…) At the time the foreign delegation visited, United Nations relief officials were warning that a long drought and a harsh winter were confronting up to a million Afghans with starvation. The envoy said that when the visitors offered money to repair and maintain the statues, the Taliban's mullahs were outraged. The mullahs told them that instead of spending money on statues, why didn't they help our children who are dying of malnutrition? They rejected that, saying, 'This money is only for statues.' 'The mullahs were so angry, they said, 'If you are destroying our future with economic sanctions, you can't care about our heritage.' And so they decided that these statues must be destroyed. The Taliban's Supreme Court confirmed the edict"

---

[3] There are somewhat different versions of this decree quoted in Western scientific literature. We prefer the version of Thomas Ruttig, founder of the Afghan Analysts Network (AAN), because it relies directly on the Pashtu source (Ruttig 2011). Other versions of this notorious decree can be found for ex. in (Falser 2011, pp. 159f; Van Krieken-Peters 2006; p. 210; Francioni and Lenzerini 2003, p. 626) and in several other publications. It should be noted that Mullah Omar's decree has never been published in the legal Gazette in Afghanistan, and has only been broadcasted on Radio Shari'at or appeared on national newspapers, therefore, there does not exist an "official" version.

[4] A circumstantiated chronology of the events can be found in Elias (2007, pp. 17–19) and Butt (2001, pp. 53–56).

(Crossette 2001). The envoy further stressed that the destruction of the Bamiyan Buddha statues must not be understood as a measure of religious repression, since there are no Buddhists in Afghanistan, whereas Hindu temples in the country would continue to be protected. Around the same time, Mullah Omar insisted in a statement that he preferred to be remembered as a "smasher of idols (*bot-shekan*) rather than a seller of idols".[5]

The demolition of the Buddha statues was so carefully orchestrated for the international news that its outreach should chiefly be analyzed in the international community. Nationally and regionally, public reaction was quite weak. The Buddha statues can hardly be addressed as symbols of Afghan cultural identity,[6] but they had formerly gained some prominence on a national level, having been used on banknotes in 1939 (reproduced in Morgan (2012, p. 32) and on postal stamps in 1979 (reproduced in Elias (2007, p. 14).[7] Internally, their destruction did not stir an immediate outcry,[8] since both public opinion and the press were under strict control. Only after the fall of the Taliban regime did poems and short stories about the destruction of the Buddhas appear in Afghanistan.[9] The local Hazara population, who had lived among the Buddha statues for centuries, and who had integrated them into the local folklore (Morgan 2012, pp. 129f; Hackin and Kohzad 1953; Klimburg-Salter 2020), was certainly opposed to their destruction as it was opposed to the Taliban regime more generally, but statements about the Buddha statues as incarnations of Hazara cultural identity appeared only after the fall of the Taliban Emirate (for ex. (Ahmadi 2012, pp. 51f)).

It makes no sense to examine the censored Afghan media of that time, but largely enthusiastic reactions in the Pakistani press have been described by Jamal J. Elias (Elias 2007, pp. 22ff). The same author indicated that the demolition of the Bamiyan Buddhas occurred during the Haj period of the Islamic calendar (Elias 2007, p. 20) around the Eid-al Adha holiday which commemorates Abraham's sacrifice of Isaac, thus underlining the strictly monotheistic creed of Islam. It is evident that the timing of the event was carefully planned, charging it with religious significance. The Taliban

---

[5] This phrase is a quotation of Mahmud of Ghazni who in 1025 plundered a Shiva temple on his way to conquer India (Flood 2002, pp. 651f).

[6] Pierre Centlivres chiefly blames work of the *Délégation Archéologique Française en Afghanistan* (DAFA) for the alienation of the Bamiyan (and generally Afghan) population from "their" Buddhas because it established the "Buddhist" identity of the site and claimed a Hellenistic provenance starting the eviction of the population from the inhabited caves. He also quotes various derogatory remarks from DAFA members about the local population (Centlivres 2001).

[7] Centlivres (2009, p. 16) gives different date specifications.

[8] Michael Semple, who at that time was United Nations Regional Coordination Officer in Afghanistan, in 2011 reported on *Afghan Analysts Network* (AAN) that many Taliban officials were secretly objecting to the decision to destroy the Buddhas (Semple 2011).

[9] An overview is furnished in (Haldar (2012, pp. 55–62). The 2011 short story "The Idol's Dust" by Zalmay Babakohi is easily accessible in translated form online: https://wordswithoutborders.org/read/article/2011-05/the-idols-dust/. The same stands for the poem "The Fallen Buddha Statues in Bamiyan" (2012) by Lina Rozbih-Haidari: https://www.diplomaticourier.com/posts/the-fallen-buddha-statues-in-bamiyan.

Emirate certainly did not see a Buddhist danger in Afghanistan, yet it is clear that the regime saw fit to ruthlessly reaffirm its own monotheism.[10]

Western observers agree that the blowing up of the Bamiyan statues was mainly orchestrated as an international media event, following a "modern" logic, directed mainly towards an international, i.e. Western, public (Falser 2011, p. 163; Francioni and Lenzerini 2003, p. 620; Flood 2002, p. 651; Morgan 2012, pp. 20–23; Boggs 2017, p. 30). In its immediate reaction in the *Heritage at Risk Report* 2001, ICOMOS declared that the demolition "would remain written in the pages of history as among the most infamous acts of barbarity" (ICOMOS 2001, p. 26). The term "vandalism" with which the destruction has frequently been labeled in the West, together with "barbarity", does not seem fully appropriate, since acts of vandalism are not ordinarily premeditated. In the following years, the concept of "iconoclasm" has largely prevailed in the scientific literature ((Falser 2011; Flood 2002; Morgan 2012) among others) and this is for good reason: Iconoclasm not only has a long history behind it (the same can be said for both vandalism and barbarity), but also has a theory behind it.[11] It is, in this sense, both deliberate and elaborately staged.[12]

We cannot enter into debates about what the Taliban really intended with their iconoclastic policies, and which theory stood behind their specific iconoclasm. The opinion of some commentators that the so-called "idolatry" that the Taliban wanted to eradicate, evidently not the Buddhist one, should consist of the aesthetic, western idolatry of heritage itself (Falser 2011; Crossette 2001; Klimburg-Salter 2020) seems too far-fetched and confers too much cultural credit to the Taliban Emirate. These commentators see in Taliban iconoclasm a "revolt against the museum" itself and hold that the status of the destroyed objects as world heritage not only did not shield them, but rendered them a target (Latour 2002). Such hermeneutics of the Taliban mind do not lie within the scope of the present contribution; suffice it here to say that in 2001 the Bamiyan Buddhas were still not considered world heritage. Since such interpretations are not substantially supported by the available documents, they appear dictated ultimately by the urgency of Western critical intellectuals to distinguish themselves through bodacious paradoxes. Given what we currently know, we have to limit ourselves to view the iconoclastic Taliban policies as being little more than a provocative act of self-affirmation; an attempt to assert their sovereignty over the country, including its cultural heritage. In this sense, the message has indeed been directed towards the West.

---

[10] Francioni and Lenzerini speak about an "act of narcissistic self-assertion" (Francioni and Lenzerini 2003, p. 621).

[11] The obligatory analysis of iconoclasm in early Christian contexts, especially in the Eastern Churches, and in early Protestant movements remains Warnke (1973). An overview over iconoclastic tendencies in Islam is offered in Flood (2002).

[12] Michael Falser has used the term "performative iconoclasm" for the Bamiyan case (Falser 2011); "performative", of course, in the sense of John Austin's speech-act theory, in which "performative acts" are those that realize their significance through their very performance (like, for ex., the wedding set phrases). On the other hand, Bruno Latour's term "iconoclash", to which Michael Falser refers, does not seem to fit the Bamiyan case, because it implies the clash of two conflicting iconic spheres (Latour 2002).

In 2001, Afghanistan still had no site on the UNESCO World Heritage list, but the Taliban demolition of the Buddha statues paved the way for the emergency nomination of two World Heritage sites in the country: first, the Minar-e Jam from the Islamic (Ghorid) period and, a few months later, the "Cultural Landscape and Archeological Remains of the Bamiyan Valley" (2003).

3 Bamiyan as World Heritage Site

Both UNESCO and its advisory body ICOMOS reacted immediately and on a high-ranking diplomatic level to the events occurring in Afghanistan. UNESCO director-general, Koichiro Matsuura, wrote a letter to Mullah Omar on the 28th of February and, on the 8th of March 2001, and convinced the President of Egypt, Hosni Mubarak, to send a delegation of Al-Azhar scholars, the most prestigious religious institution in the Sunni world, to Kandahar. The delegation arrived on the 11th of March, while the demolition of the Bamiyan Buddhas was underway (Bouchenaki 2020, pp. 19–30). The Al-Azhar delegation was accompanied by Pierre Lafrance as a special UNESCO envoy, and by Yusuf al-Qaradawi, the most popular Arab preacher (Elias 2007, p. 17) of the time. On the 12th of March, having been informed that all efforts had been in vain, UNESCO's director-general labeled the demolition of the Bamiyan Buddhas and other iconoclastic acts carried out by the Taliban government, "a crime against culture",[13] as per the terms of the *Hague Convention for the Protection of Cultural Property in the Event of Armed Conflict* of 1954, and that it ought to be taken to the International Court of The Hague.

Once the destruction of the Bamiyan statues was accomplished, UNESCOs World Heritage Committee was quick to organize the first safeguarding measures to protect what had survived.[14] In January 2002, UNESCO was officially requested by the Minister of Foreign Affairs of the Afghan interim administration to play a coordinating role in all future international and bilateral activities aimed at safeguarding Afghanistan's cultural heritage. A Memorandum of Understanding was signed in March 2002 with the Afghan Minister of Information and Culture, Mr. Said Makhdoom Raheen, entrusting UNESCO with the coordination of international efforts for the restauration of the National Museum of Kabul. In May 2002, UNESCO organized a seminar in Kabul, continued at a conference in Tokyo held in July 2002 on "The Culture of Afghanistan", where Mr. Matsuura personally handed over to Minister Raheen the certificate of inscription of the Minaret of Jam on the World Heritage List. He also announced that at the next session of the World Heritage Committee, the proposal to include the Bamiyan site on the World Heritage List would be considered (UNESCO dossier DG/2003/086). The international meetings

---

[13] For the concept of "crime against culture" and the connected stipulations in international law, see Francioni and Lenzerini (2003) and Prott (2006), Maniscalco (2006).

[14] The following data is extracted from (Manhart 2002; Manhart 2003; Manhart 2004; Manhart and Lin 2014, pp. 61–67; LaGeS 2018, pp. 39–49).

were also a chance to install a (short-lived) International Coordination Committee for the Safeguarding of Afghanistan's Cultural Heritage (ICC) at the UNESCO headquarters in Paris, soon supplanted by the Bamiyan Experts Working Group (BEWG), organized by UNESCO and ICOMOS, which held its first meeting in Munich (Germany) from 21 to 22 November 2002. The BEWG had then met 14 times by 2017 and had established the framework, which would bring together international conservation experts and representatives of the Afghan government and the principal donor countries (Japan, Italy, Germany, South Korea e.a.). In fact, spanning 2002–03, more than 7 Mio $ could be raised for Bamiyan alone (Manhart 2003, p. 83).

In December 2001, UNESCO organized a mission to Bamiyan with the aim of assessing the condition of the site, while covering the remaining rock fragments with fiberglass sheets, which would protect them from harsh weather conditions during winter. In July 2002, a second UNESCO mission, jointly organized with ICOMOS, was undertaken so that conservation measures could be prepared at the Bamiyan site. A third ICOMOS mission, aimed at further substantiating the project and composed of German, Italian and Japanese experts was then undertaken from the 27th of September to the 6th of October 2002. These missions laid the groundwork for the nomination of the "Cultural Landscape and Archaeological Remains of the Bamiyan Valley (Afghanistan)" as a World Heritage site (World Heritage Committee, Decision 27 COM 8C.43) on the 5th of July 2003.[15] Simultaneously, Bamiyan was placed on the cultural heritage "in danger" list, as had already happened with Minar-e Jam. Bamiyan's nomination bears all the marks of an emergency nomination, made hastily in order to take a stand against Taliban iconoclasm. Bamiyan, in fact, was nominated as a "cultural"—not "natural" or "mixed"—heritage site, more specifically as a "serial nomination", naming eight quite distant and historically heterogeneous archeological sites in the valley, among which the central Buddha cliff, which came in the first place, plus a "cultural landscape", not defined topographically or conceptually.

The nomination of Bamiyan by UNESCO (dossier 208rev), laid the foundations for the intense conservation work in the Buddha niches, the surrounding caves and other archeological sites that were to follow over the next years. In this context, the question of a possible recovery of the Buddha statues has been debated right from the beginning. The Japanese painter and scholar Ikuo Hirayama suggested leaving the Buddha niches empty immediately, as a powerful monument in their own right (Morgan 2012, p. 199). Other proposals were of more questionable taste. On the 15 July 2002 issue of the *New Yorker*, the artist Otto J. Seibold responded to an invitation to design a fitting memorial of the site of the Twin Towers destroyed on 9/11: his image depicted the two Buddha sculptures of Bamiyan recreated in Manhattan, while

---

[15] On June 16, 2003, ICOMOS had already published a dossier, no 208 rev, which became the basis of the UNESCO nomination document, https://openarchive.icomos.org/id/eprint/437/1/af-208rev. pdf.

miniature Twin Towers, housing refugees, were presented as occupying the empty Bamiyan niches.[16]

Apart from such individual proposals, the international heritage conservation community answered to the reconstruction demand from the Afghan state party in the above mentioned seminar on Afghan Cultural Heritage, May 27–29, 2002, in Kabul (conclusions IV/14): "The decision to engage in the reconstruction of the Buddhist statues of Bamiyan is a matter to be settled by the Government and people in Afghanistan." For the moment, the international experts would postpone the issue: "The Seminar participants underscored that such work could be undertaken only after major stabilization work on the cliffs at Bamiyan has been completed" (Toubekis et al. 2017, p. 41). This deferral of an admittedly sensitive, complex and potentially very expensive decision set the tone of the debate for the following years, whose general pattern opposed the skepticism of international conservation experts against Afghan demands (Morgan 2012, p. 180; Toubekis et al. 2017, p. 274). For the time being, ICOMOS would concentrate on "Safeguarding the remains" of the Bamiyan statues—as the subtitle of voluminous technical documentaries of their works states (ICOMOS XIX 2009; ICOMOS XXI 2016)—rather than on any form of reconstruction. On the 9th BEWG meeting of 2011, the president of ICOMOS international, Michael Petzet, declared: "I believe it will be no disaster if not all the dreams of a total 'resurrection' of the famous Bamiyan Buddhas … can be implemented here and now. Considering the disaster of ten years ago we can be quite satisfied with having saved what could be saved—in the awareness that our efforts could open up chances for future generations to continue working in the sense of … partial reconstruction" (ICOMOS XXI 2016, p. 93, again 2012, p. 180). This variously repeated statement was based on the fact that a full reconstruction would imply "reconstructing" a statue of which important details (face,[17] arms and hand positions, feet) are unknown.[18]

Moreover, managing the remains of the Buddha statues would have had to resolve a series of problems:

(1) How to conserve the highly fragile rock fragments that had remained and been recovered.
(2) How to display these remains, given that their placement in a museum is ruled out by their sheer mass.
(3) Finally, how to restore some idea of the Buddha figure that once inhabited the two niches.

The ICOMOS experts, and especially its international president, Michael Petzet, the "leading champion of anastylosis", as Morgan calls him (Morgan 2012, pp. 199f),

---

[16] The images are reproduced in Morgan (2012, pp. 24f).

[17] The missing face of the Buddha statues above the upper lips is not due to previous Islamic iconoclasm. The *Indian Archeological Survey*, which worked in Bamiyan from 1969 to 1974, has made it highly probable that the missing face was substituted by a—painted and decorated—wooden mask for which appliances were discovered (Sengupta et al. 2008).

[18] A complete reconstruction would thus be equivalent to a speculative "stylistic restauration" à la Viollet le Duc, anathema in Western conservation circles since the *Venice Charter* of 1964 (Jokilehto 2018, pp. 183ff).

therefore opted right from the beginning not so much for a "reconstruction", but for an anastylosis, in which the original fragments of the statues would be replaced in their original position, within a steel structure roughly the shape of the original Buddha, within the niches. This was the proposal pursued right from the beginning. The first BEWG meeting of 2002 in Munich already stated: "The experts welcome that the Afghan authorities acknowledge the possibility of an anastylosis as one well-established method of proper relocation of the rock fragments to their original position." According to its supporters, this solution would satisfy the three requirements listed above, would conform to §§ 9 and 15 of the *Venice Charter* and to § 86 of UNESCO's *Operational Guidelines for the Implementation of the World Heritage Convention*, and would, furthermore, "preserve the authentic spirit of the place" (ICOMOS XIX 2009, p. 14). In 2002, Petzet published an essay, *Anastylosis or Reconstruction: The Conservation Concept for the Remains of the Buddhas of Bamiyan*, in which he attempted to settle the question on theoretical grounds. In particular, he contrasts his proposal of anastylosis with competing ones and states:

(1) A virtual reconstruction of the statues, proposed and prepared by Armin Grün (Zürich) would not resolve the problem of the management and preservation of the remaining, original rock fragments (Grün et al. 2004), see also Toubekis et al. (2017).

(2) The same holds for a reconstruction of one or both Buddhas using traditional techniques, suggested by an Afghan sculptor, i.e. hewn out of the rock and coated with loam plaster. This solution would, furthermore, irretrievably destroy what has remained of the original statues on the back walls of the niches. Deepening the niches by a few meters would also compromise their connection with the surrounding caves.

(3) "A reconstruction with modern materials (a brand new Buddha made of concrete?) or at least its evocation with laser techniques in a future sound-and-light show seems under the present circumstances rather strange" (ICOMOS XIX 2009, pp. 46–48).

In order to circumvent the restrictions placed on reconstructions in both the *Venice Charter* and UNESCO's *Operational Guidelines*, Petzet makes reference to the Old City of Warsaw, the Mostar Bridge, the Xian Terracotta Army and many other examples in which the international conservation community has accepted even a complete reconstruction.[19] An anastylosis, in short, requires a "provisional structure" within the Buddha niches, made out of steel or another load-bearing but inconspicuous material, into which the original rock fragments would be put, in order to be integrated in their original position. Anastylosis, so the argument goes, would at the same time resolve the problem of the reconstitution of the Buddha statues, and that of where and how to place the remaining original rock fragments. However, according to Petzet, the limits of anastylosis would be reached "when only a few original fragments would appear as a sort of 'decoration' on the provisional structure" (BEWG meeting

---

[19] In a recent contribution, Mechtild Rössler lists all international charters and all UNESCO decisions concerning the reconstruction debate (Rössler 2020).

2011, (ICOMOS XXI 2016, p. 131), see for 2014, (ICOMOS XXI 2016, p. 246)).[20] The final decision, Petzet concludes, lies in any case with the Afghan government (ICOMOS XXI 2016, p. 140). Yet, even the hypothesis of anastylosis, on which ICOMOS has more or less committed itself,[21] should not be implemented without issue, because it requires continued investigations regarding the technical possibility (BEWG meeting 2006, recommendations II/10, (ICOMOS XIX 2009, p. 125).[22]

In the meantime, the Afghan side insisted on the rapid carrying out of said anastylosis; in 2009, Habiba Sarabi, provincial governor of Bamiyan, in her foreword to the ICOMOS documentation, wrote: "One of the key achievements of ICOMOS has been in terms of raising hope among Afghans for rebuilding the Buddhas, at least one of the statues, by using the remaining original pieces and with external materials to exhibit it as one of the memorial monuments of the historic past and as witness to the cultural journey of glory and of painful suffering, including the destruction of the Buddhas. I strongly support the idea and initiatives to restore this rich cultural heritage" (ICOMOS XIX 2009, p. 11). Again, in 2013, the Afghan Minister of Information and Culture stated in a BEWG meeting: "There is still strong support in Afghanistan for the reconstruction of at least one of the Buddha sculptures destroyed by the Taliban in 2001. This can be a symbol for Afghanistan and the World that the new Afghanistan is ready to stand for peace, democracy and an open society against fanaticism" (ICOMOS XXI 2016, p. 197). The Minister concluded his statement by expressing his bewilderment at the international worrywarts.[23] But even though the Afghan side expressed a desire for some kind of "reconstruction", it was not clear which form was to be favored, as the ICOMOS experts repeatedly observed (ICOMOS XXI 2016, p. 240).

An ICOMOS meeting in 2014 in Tokyo went a step further. The conservation experts concentrated on the eastern, smaller Buddha niche, which had successfully been stabilized, and was brought into the discussion regarding the possibility of its "reconstruction", its "partial reconstruction", its "reassembly", finding a way around using the term "anastylosis". However, they came across another obstacle, as it was noted by the Japanese delegates, that the "deliberate destruction" of the Buddha statues was explicitly mentioned in the definition of the Outstanding Universal Value of the site in the UNESCO nomination dossier (criterion VI). This meant that any

---

[20] It is evident, however, that "the roughly 9.000 plaster fragments salvaged from the rubble cannot be repositioned, they will be important documents in the archive of findings for future research" (ICOMOS XXI 2016, p. 246).

[21] Also the term "partly anastylosis" appears in the ICOMOS reports (ICOMOS XXI 2016, pp. 17, 59, 130f). An alternative briefly (2010–11) considered, was to display the conserved rock fragments of the statues in a lapidarium (see ICOMOS XXI (2016, p. 73)).

[22] More or less verbatim repeated at the BEWG meeting, 2008 (ICOMOS XIX 2009, p. 136), 2012 (ICOMOS XXI 2016, p. 176).

[23] This feeling was evidently shared by large parts of the Bamiyan population. One member of the ICOMOS team was recorded saying in 2015: "Local employees of international heritage organisations reported being constantly asked by residents about the rebuilding and when it will happen. Many of my informants were frustrated at a perceived lack of action at the Buddha niches fourteen years after the destruction, asking 'Why don't they just rebuild'?" (ICOMOS XXI 2016, p. 394).

reassembly of the empty Buddha niches that could affect the site's Outstanding Universal Value had to first be discussed (ICOMOS XXI 2016, pp. 216f).[24]

Furthermore, several, purely technical questions had not been unanimously answered in these discussions:

(1) What is the percentage of original sculpture fragments to be displayed in an anastylosis for each Buddha?
(2) Which of the two Buddha niches (or both) should be considered for reconstruction?
(3) Are the rock fragments stable enough to be fixed and displayed in a steel construction?
(4) Most importantly, what are the estimated costs?

On the question of how much of the original, salvaged material should be displayed in an anastylosis, the scientific community did not reach a consensus. ICOMOS expert Bert Praxenthaler estimated that for the eastern Buddha, roughly 40–45% of its mass had been salvaged, of which 90%, which constitutes approximately 300 stone fragments, could be repositioned in the niche (ICOMOS XXI 2016, pp. 36 and 231f). The ICOMOS experts from RWTH Aachen (Germany), however, estimated the useable material for the eastern Buddha as being just 12% of the original mass of the statue (Toubekis et al. 2017, p. 275 and Han et al. 2018, p. 43). The experts agree, if nothing else, on the fact that the bigger, western Buddha, made up of blocks weighing up to several tons, is better suited than the smaller, eastern one, for a possible "reconstruction". The reconstruction debate had concentrated on the eastern Buddha simply because its niche had been successfully stabilized, whereas the works on the western niche are not completed even until now (Lawler 2003). Experts also agree that the stone blocks, consisting of fragile conglomerate material, cannot simply be placed in a supportive steel structure, but need previous stabilization through an ether resin treatment in a vacuum chamber which, according to some experts, was only possible in Germany (ICOMOS XXI 2016, p. 76).

The costs for the preliminary work of ordering and sorting the salvaged fragments were estimated in 2002 at 1,5 Mio $ (Lawler 2003). The costs of a possible reconstruction-anastylosis of the statues, of course, enter an entirely different realm. The only estimate to be found in the voluminous ICOMOS documentation was provided by an ICOMOS Mission in 2014, which calculated the costs at about 30 Mio $ per statue (ICOMOS XXI 2016, p. 232). Other authors, not directly involved in this UNESCO debate, indicate even higher amounts.[25]

In sum, fifteen years of international endeavors in Bamiyan have certainly produced important achievements (the definite stabilization of the eastern niche,

---

[24] In these discussions, it became evident, that especially the Japanese delegation was strongly skeptical towards any reconstruction hypothesis (see ICOMOS XXI (2016, pp. 220f)). The discussion between German and Japanese conservations experts on the "reconstruction" of the eastern Buddha, triggered by a supposed reconstruction of its feet (2014), is copiously documented in ICOMOS XXI (2016, pp. 621–653).

[25] Centlivres in (2009), estimated the "rebuilding" (and not the anastylosis) costs for the bigger, western Buddha "en béton special" between 40 and 60 Mio $ (Centlivres 2009, p. 12).

the ongoing stabilization of the western niche, the protection of the surviving wall paintings etc.) (Praxenthaler and Beckh 2020). However, the question of a reconstruction, anastylosis, partial anastylosis, reassembly or some other version of physical reconstitution of the Buddha statues has remained open. Important technical issues are still being debated and, most importantly, the question of funding persists. Yet, the question of reconstruction has been the one that the Afghan side and the general public have shown the most interest in, because the stabilization measures and other restauration works remain, in a certain sense, invisible to non-expert eyes.

4   Three Proposals for Reconstruction

Finally, UNESCO called for a special meeting to take place in Tokyo, from the 27th to the 29th of September 2017, with the title *The Future of the Bamiyan Buddha Statues*: *Technical Considerations and Potential Effects on Authenticity and Outstanding Universal Value*. This was immediately followed by the 14th meeting of the BEWG (1–2 October). Most of the contributions were published by UNESCO in 2020 under the title *The Future of the Bamiyan Buddha Statues*. The title of the conference and of the publication refers to both Buddha statues, although, at UNESCO's request, the technical proposals were focused on the smaller, eastern Buddha niche, which had already been stabilized. Hereby UNESCO followed a request from the Afghan government on the 40th session of the World Heritage Committee in Istanbul (2016) that at least one of the Buddha statues be reconstructed. The bigger, Western niche should, for the time being, be left empty as a lasting memory of its tragic destruction. The conference volume does not, however, include ICOMOS's plans for anastylosis or partial reconstruction presented at the Tokyo conference by Bert Praxenthaler (ICOMOS Germany), but stages three new ideas, which had not yet appeared in the debate.

The arena for the three new proposals was laid out by J. Okahashi's essay, showing that Bamiyan's case fulfills all "exceptional circumstances" in which the UNESCO *Operational Guidelines for the Implementation of the World Heritage Convention* (§ 86) allows for a reconstruction (Okahashi 2020). We will briefly present and assess the three proposals.

(1) The first technical proposal was brought forward by an Italian team consisting of Claudio Margottini, Andrea Bruno e.a. (Margottini et al. 2020). The proposal ultimately aims at a comprehensive revitalization of the entire "southern branch of the Silk Road (Margottini et al. 2020, p. 275, 278), thus including Herat, Minar-e Jam and potentially many more sites. But the Tokyo conference contribution is focused solely on Bamiyan.

The authors criticize the previously dominant hypothesis of anastylosis as "impossible" (Margottini et al. 2020, p. 277, 284) in the case of the eastern Buddha, because the salvaged fragments are too few and too precarious. Furthermore, UNESCO has recognized the destroyed statues as part of the Outstanding Universal Value of the site (Margottini et al. 2020, p. 284) and not the reconstructed ones. Upon briefly

Fig. 1 Image courtesy of Margottini et al. (2020, p. 291, fig. 17)

considering the possibility of leaving the niches empty or simply putting a photo of the original statues on fiberglass panels inside them (Margottini et al. 2020, pp. 287f),[26] the authors then bring up their true proposal which they rightly call "revolutionary" (Margottini et al. 2020, p. 304). Drawing on the famous thesis of the "Technical Reproducibility of the Work of Art" in our times (Margottini et al. 2020, p. 277), the authors broach the possibility of filling the eastern Buddha niche with an exact replica of the statue, produced with digital robotic techniques, in shining white Carrara marble. To reduce the weight, of course, these statues would consist only of a 10–15 cm marble shell, supported by an internal steel structure. Since the marble would be extracted from the same quarry already used by Michelangelo, it would both connect Bamiyan with the renaissance artist while "conveying to Afghanistan a message of optimism. (…) The Buddha's expressive eyes will find, in this marble material, a new light" (Margottini et al. 2020, pp. 288–290). Indeed, such a proposal respects two fundamental criteria of any modern reconstruction. It is reversible and thus not prone to future terrorist attacks and it is clearly recognizable by its original parts (Fig. 1).

Directly in front of the Buddha niche, a museum should be constructed underground, with a loophole on top, through which to observe the Buddha statue above (Margottini et al. 2020, p. 284). This museum should be situated underground in

---

[26] This solution was implemented temporarily in 2002 by Margottini et al. (2020, p. 277).

order not to disturb the landscape. Such a museum might infringe, however, on the protected archeological area, especially in front of the eastern Buddha where no archeological survey has ever been conducted. The rock fragments of the original statue temporarily contained in a shack in front of the niches, on their side, do disturb the landscape (in the opinion of the authors) and should be placed elsewhere; the question of where and how remains unresolved.

On the Bamiyan Valley's southern fringe, opposite the Buddha Cliff, the proposal outlines a plan for a museum equipped with all refinements of Virtual and Augmented Reality (Margottini et al. 2020, p. 282). The authors do not take into account that exactly in this place the Bamiyan Cultural Centre stands, which was inaugurated in 2020, and whose construction began in 2016. Instead, the proposal resubmits the project by architect Andrea Bruno, which was not considered in UNESCO's international bidding in 2015 (Margottini et al. 2020, p. 280). Finally, the proposal does not contain any expense budgeting.

(2) The second technical proposal was put forward by Michael Jansen, Georgios Toubekis and Matthias Jarke from the Technical University (RWTH) of Aachen/ Germany (Toubekis et al. 2020). The authors cater to an "authentic remodeling of the Eastern Buddha with the integration of fragments", which should be "community-based" and follow the "traditional building techniques with the intention of retaining tangible and intangible values" (Toubekis et al. 2020, p. 307). Such an "authentic reconstruction", using for a moment this oxymoron, should be part of a "necessary reconciliatory process",[27] "save the spirit of the place" (Toubekis et al. 2020, p. 308 and 313) with an "authentic piece of artwork" (Toubekis et al. 2020, p. 322) and, furthermore, contribute to "forge a national identity" for Afghanistan (Toubekis et al. 2020, 312 and 326).

Whereas the fragments of the bigger, western Buddha should be spread out horizontally in front of the empty niche (Toubekis et al. 2020, p. 311), the eastern Buddha should be reconstructed by the local people themselves (Toubekis et al. 2020, p. 310) using the most authentic local building material available, sun-dried clay. "Mud has been used as raw material on the figures in previous times, and it originates from the Cultural Landscape. Thus it can be considered an authentic material, with respect to the Outstanding Universal Value of the property" (Toubekis et al. 2020, p. 322). Into this statue of mud, the salvaged original fragments, about 12% of the total mass according to the authors (Toubekis et al. 2020, p. 316), should be embedded. In order to sustain a sculpture of 38 m, the mud has to be reinforced somehow. In fact, the authors ultimately discourage a precise copy of the original stone statue being made, because protruding parts like arms cannot be modeled using mud (Toubekis et al. 2020, p. 317).

In addition to this reconstruction, the authors envision Augmented Reality applications via smartphones that accompany visitors on the site. The Bamiyan Cultural Centre in front of the Buddha Cliff should host "an interpretation and mediation section that is able to explain the history of the Bamiyan site and of its research and

---

[27] The issue of heritage management as reconciliation has only recently entered into the UNESCO debates (Ringbeck et al. 2022, pp. 439–444).

documentation over time" (Toubekis et al. 2020, p. 310). How the mediation and interpretation of the site might be achieved, is not further explained. The authors abstain from providing an estimate of cost. They also do not furnish a rendering of the reconstructed Buddha statue.

(3) The third technical proposal was presented at the Tokyo conference by a team of nine Japanese experts (Maeda et al. 2020). This project relies fundamentally on the idea of leaving the site of the empty Buddha niche itself untouched. The authors repeatedly warn against any "premature reproduction" (Maeda et al. 2020, p. 333) and opt for keeping to the status quo, a principle of prudence which probably owes something to the prominence of archeologists in the Japanese team, who recognize that "the tragic destruction in 2001 itself is considered invaluable to The Property as a World Cultural Heritage" (Maeda et al. 2020, p. 335).

The suggested reconstruction of the Buddha statue should, therefore, take place outside the protected area of the UNESCO property, precisely on the southern slope of the valley, opposite the empty Buddha niche. The authors propose a reduced, 13 m replica of the Eastern Buddha, made of fiberglass reinforced concrete (Maeda et al. 2020, p. 345). It should be accessible through a stairway which would also lead to newly hewn caves, and which would host a museum. Only the rock fragments of the original statues should be stored near their niches (Maeda et al. 2020, p. 337). The cost of the entire endeavor is estimated at 13,34 Mio $. (Maeda et al. 2020, p. 346).[28]

The authors do not mention that the site of the reduced new Buddha and the attached museum would be exactly where the Bamiyan Cultural Centre was under construction at that time. Nor do they integrate it into their proposal. They only refer to a long-abandoned Japanese project for the Cultural Centre, presented at the 12th meeting of the BEWG in 2013 (Maeda et al. 2020, pp. 336f).

All three Tokyo proposals seem problematic, aside from possible objections on aesthetic grounds, because they do not sufficiently take into account the local context. The proposals put forward by Margottini and Bruno as well as that of the Japanese, fail to acknowledge the existence of the Bamiyan Cultural Centre. Furthermore, the Margottini and Bruno project does not indicate how best to manage the salvaged rock fragments of the sculptures. The Japanese proposal, on the other hand, underlines the safeguarding of the archeological areas, which has been more or less ignored in the proposals by both Aachen and Margottini/Bruno. Lastly, the Aachen proposal does not consider the weight of an entire Buddha statue made entirely out of mud, which would greatly burden the ground under the niche.

However, no decision regarding the three proposals was made at the Tokyo meeting in 2017. Instead, the subsequent BEWG meeting "recommended that the Afghan authorities establish a committee to review and discuss the submitted proposals, and any further proposals, to identify the future action to be taken for the Buddha niches (V.3). Thus, the meeting ended with a further deferral. Given the political turmoil in Afghanistan, this committee was never established, and the issue of Bamiyan's giant Buddha statues remains unresolved.

---

[28] Unfortunately, we have not received the permission to reproduce a rendering of the Japanese project in this volume.

5 Some Conclusions

The events of August 2021 interrupted the long debate surrounding how a reconstruction of Bamiyan's Buddha statues might eventually take place. In this new scenario of the Islamic Emirate, any attempt at reconstruction, revivification, reassembly, virtual or physical, is out of question. This is not merely due to the new government's presumed opposition to it, but also for the all-too-evident economic hardships faced by the population, to which such a costly endeavor would stand in unsustainable contrast. However, this does not unburden us from the task of trying to find a reasonable solution to settle this two-decade-long debate.

The proceedings of the 2017 Tokyo conference also contain a contribution by Jukka Jokilehto, head of ICCROM, another of UNESCO's advisory institutions, which, in our opinion, could serve as a stimulus for further consideration. Jukilehto speaks out against anastylosis, because the rock fragments that have to be relocated in the empty niche "are just material without form" (Jokilehto 2020, p. 212). But Jokilehto also reintroduces an argument that had already been made in the UNESCO debates: any form of reconstruction would necessarily render invisible and obfuscate some part of what has remained. It would introduce a new and overwhelmingly prominent element into Bamiyan's cultural landscape, whose Outstanding Universal Value was recognized in 2003 with the empty niches as part of it. Jokilehto concludes: "It is not advisable to propose any reconstruction or anastylosis in the ancient niches. The present remains are the most efficient memorial to the 2001 destruction, and they are the most authentic and prestigious monument for the history of the Bamiyan Valley and its community" (Jokilehto 2020, p. 213).[29]

Following the highly publicized destruction of the Buddha statues in 2001, any form of physical reconstruction appears gratuitous today, since the image of the Buddha statues, as well as the image of their destruction, are virtually omnipresent in the media. Every visitor in front of the empty Buddha niche already has the image of the Buddha impressed in his or her mind. A material replica would thus simply be redundant.[30]

In the Tokyo conference volume, James Janowski argued in favor of reconstruction, on philosophical grounds.[31] The original Buddha statues are a testament to mankind's achievements and should not be lightheartedly be given up, but which

---

[29] Deborah Klimburg-Salter, the leading historian of the "Kingdrom of Bamiyan", largely takes up these arguments (Klimburg-Salter 2020, p. 234).

[30] In 2013, Daniel Fabre coined the concept of "émotions patrimoniales" which have to be preserved beyond their material support (Fabre 2013). These emotions are certainly more powerfully stirred up by the empty niches because the imagination superposes the image of the statues with the image of their explosion.

[31] If we want to enter into a philosophical dimension, one could see the debate around the Bamiyan Buddha statues as an ironic instantiation of Jacques Derrida's famous *différance*: in the end, it has been the continuous deferral that has made the difference (Derrida 1967).

can be revived by the new achievement of reconstruction (Janowski 2020, pp. 257–273).[32] This might be true, but in our view, more importantly, the destruction of Bamiyan's Buddha statues was a historical disaster and historical disasters cannot be redeemed nor repaired. They must be remembered and taken into account.

References

Ahmadi R (2012) Bamiyan Buddhas: view from Hazara. Himal Cent Asian Stud 16(2):51–52. http://www.himalayanresearch.org/pdf/2012/vol16n2.pdf. Last Accessed 19 Mar 2020. (Bamiyan Special)

Boggs E (2017) UNESCO takes on the Taliban, the fight to save the Buddhas at Bamiyan. Va Undergrad Hist Rev 5:23–33. https://doi.org/10.21061/vtuhr.v5i1.39

Bouchenaki M (2020) Safeguarding the Buddha statues in Bamiyan and the sustainable protection of Afghan cultural heritage. In: Nagaoka M (ed) The future of the Bamiyan Buddha statues. Heritage reconstruction in theory and practice, pp 19–30. Springer, UNESCO

Butt JM (2001) The Buddhas of Bamiyan: saving other possible cultural targets. In: OIC, UNESCO, Doha conference of Ulama on Islam and Cultural Heritage, Doha 30–31 December 2001, pp 53–56. https://unesdoc.unesco.org/search/7d8b2451-52e8-48d1-92be-0947d95c4c56. Last Accessed 02 May 2011

Cassar B, García R (2006) The society for the preservation of Afghanistan's cultural heritage: an overview of activities since 1994. In: Van Krieken-Peters J (ed) Art and archeology of Afghanistan. Its fall and survival, handbook of oriental studies, section eight: Central Asia, vol XIV, pp 15–38. Brill, Leiden-Boston

Centlivres P (2001) Les Bouddhas d'Afghanistan. L'épopée des archéologues, la Crosière jaune à Bamiyan, la découverte d'un site fabuleux. Éditions Favre, Lausanne

Centlivres P (2009) Vie, mort et survie des Bouddhas de Bamiyan (Afghanistan). Livraisons D'histoire De L'architecture 17(13–26):2022. https://doi.org/10.4000/lha.200.RetrievedonDecember14

Crossette B (2001) Taliban explains Buddha demolition, p 19. New York Times. https://www.google.com/search?q=10)+Crossette%2C+B.%2C+Taliban+Explains+Buddha+Demolition%2C+New+York+Times%2C+March+19%2C+2001&oq=10)%09Crossette%2C+B.%2C+Taliban+Explains+Buddha+Demolition%2C+New+York+Times%2C+March+19%2C+2001&gs_lcrp=EgZjaHJvbWUyBggAEEUYOdIBCTMzMzVqMGoxNagCALACAA&sourceid=chrome&ie=UTF-8. Last Accessed 21 Nov 2012

Darlington J (2020) Fake heritage: why we rebuild monuments. Yale UP, New Haven, London

Derrida J (1967) L'écriture et la différence. Tel Quel, Paris

Elias J (2007) (Un) making idolatry: from Mecca to Bamiyan. Future Anterior: J Hist Preserv Hist Theory Crit 4(2):12–29. www.jstor.org/stable/25835009. Last Accessed 01 Mar 2021. (JSTOR)

Fabre D (2013) Émotions patrimoniales. Éditions de la Maison des sciences de l'homme, Paris

Falser M (2011) The Bamiyan Buddhas, performative iconoclasm and the 'Image of heritage'. In: Giometti S, Tomaszewski A (eds) The image of heritage. Proceedings of the international conference of the ICOMOS international scientific committee for the theory and the philosophy of conservation and restoration, 6–8 March 2009, Florence, Italy, pp 157–170. Polistampa, Firenze

Flood FB (2002) Between cult and culture: Bamiyan, Islamic Iconoclasm, and the Museum. Art Bull LXXXIV(4):641–659. https://www.academia.edu/34934542/Bamiyan_Islamic_Iconoclasm_and_the_Museum. Last Accessed 29 Aug 2008

---

[32] An excellent and recent overview over the philosophical debate around reconstruction and authenticity is provided in Darlington (2020).

Francioni F, Lenzerini F (2003) The destruction of the Buddhas of Bamiyan and international law. Eur J Int Law 14:619–651. https://doi.org/10.1093/ejil/14.4.619

Grissmann C (2006) The Kabul Museum: its turbulent years. In: Van Krieken-Peters J (ed) Art and archeology of Afghanistan. Its fall and survival, handbook of oriental studies, section eight: Central Asia, vol XIV, pp 61–78. Brill, Leiden-Boston

Grün A, Remondino F, Zhang L (2004) Photogrammetric reconstruction of the great Buddha of Bamiyan, Afghanistan. Photogramm Rec 19(107):177–199. https://www.researchgate.net/pub lication/227635047_Photogrammetric_Reconstruction_of_the_Great_Buddha_of_Bamiyan_ Afghanistan. Last Accessed 14 Nov 2017

Hackin R, Kohzad AA (1953) Légendes et coutumes afghans. PUF, Paris

Haldar A (2012) Echoes from the empty Niche Bamiyan Buddha speaks back. Himal Cent Asian Stud 16(2):55–62. http://www.himalayanresearch.org/pdf/2012/vol16n2.pdf. Last Accessed 19 Mar 2020. (Bamiyan Special)

Han J, Bawary MR, Bruno A (2018) The Bamiyan Buddhas. Issues of reconstruction. World Herit 86:40–47. https://unesdoc.unesco.org/ark:/48223/pf0000261517. Last Accessed 04 Nov 2019

ICOMOS (2001) Heritage at risk world report 2001 on monuments and sites in danger, https:// www.icomos.org/en/what-we-do/risk-preparedness/heritage-at-risk-reports/116-english-cat egories/resources/publications/211-icomos-world-report-2001-2002-on-monuments-and-sites-in-danger. Last Accessed 04 July 2016

ICOMOS XIX (2009) The giant Buddhas of Bamiyan. In: Petzet M (ed) Safeguarding the remains. Bäßler Verlag, München

ICOMOS XXI (2016) The giant Buddhas of Bamiyan II. In: Petzet M, Emmerling E (ed) Safeguarding the remains 2010–1025. Bäßler Verlag, München

Janowski J (2020) Emptiness and authenticity at Bamiyan. In: Nagaoka M (ed) The future of the Bamiyan Buddha statues. Heritage reconstruction in theory and practice, pp 257–273. Springer-UNESCO, Paris

Jokilehto J (2018) A history of architectural conservation, 2nd edn. Routledge, London, New York

Jokilehto JI (2020) Refections on the case of Bamiyan. In: Nagaoka M (ed) The future of the Bamiyan Buddha statues. Heritage reconstruction in theory and practice, pp 205–214. Springer-UNESCO, Paris

Klimburg-Salter D (2020) Entangled narrative biographies of the colossal sculptures of Bamiyan: heroes of the mythic history of the conversion to Islam. In: Nagaoka M (ed) The future of the Bamiyan Buddha statues. Heritage reconstruction in theory and practice, pp 215–238. Springer-UNESCO, Paris

LaGeS (2018) Bamiyan strategic master plan. Polistampa, Firenze

Latour B (2002) What is iconoclash? In: Latour B, Weibel P (eds) Iconoclash, Karlsruhe, pp 14–37. http://www.bruno-latour.fr/sites/default/files/downloads/84-ICONOCLASH%20PDF.pdf. Last Accessed 29 Sept 2021

Lawler A (2003) Buddhas may stretch out, if not rise again. Science 8:298. https:/ /www.science.org/doi/https://doi.org/10.1126/science.298.5596.1204, https://doi.org/10.1126/ science.298.5596.12. Last Accessed 29 May 2022

Maeda K, Okazaki S, Sugiura N, Yamaguchi A, Miyasako M, Yamauchi K, Tamai K, Aoki S, Inoue T (2020) Technical proposal for revitalizing the Eastern Buddha statue in Bamiyan. In: Nagaoka M (ed) The future of the Bamiyan Buddha statues. Heritage reconstruction in theory and practice, pp 331–350. Springer-UNESCO, Paris

Manhart C (2002) UNESCO's response to the destruction of the statues in Bamiyan. In: Warikoo K (ed) Bamiyan. Challenge to world heritage, pp150–155. Bhavana Books, New Delhi

Manhart C (2003) UNESCO's mandate and activities for the rehabilitation of Afghanistan's cultural heritage. Mus Int 55(3–4):77–83. https://unesdoc.unesco.org/ark:/48223/pf0000133496. Last Accessed 14 Oct 2009

Manhart C (2004) UNESCO's role in the rehabilitation of Bamiyan in Afghanistan. Landslides 1:311–314. https://doi.org/10.1007/s10346-004-0033-1

Manhart C, Lin R (2014) UNESCO's activities for the safeguarding of Bamiyan. In: Margottini C (ed) After the destruction of giant Buddha statues in Bamiyan (Afghanistan) in 2001. In: A UNESCO's emergency activity for the recovering and rehabilitation of Cliff and Niches, pp 61–67. Springer, Berlin

Maniscalco F (2006) The threats to cultural heritage in the event of armed conflict: a checklist. In: Van Krieken-Peters J (ed) Art and archeology of Afghanistan. Its fall and survival, handbook of oriental studies, section eight: Central Asia, vol XIV, pp 335–352. Brill, Leiden-Boston

Margottini C, Bruno A, Casagli N, Massari G, Rüther H, Tincolini F, Tofani V (2020) The renaissance of Bamiyan (Afghanistan) and some proposals for the revitalisation of the Bamiyan Valley. In: Nagaoka M (ed) The future of the Bamiyan Buddha statues. Heritage reconstruction in theory and practice, pp 275–306. Springer-UNESCO, Paris

Morgan L (2012) The Buddhas of Bamiyan. Profile Books, London

Okahashi J (2020) Could the giant Buddha statues of Bamiyan be considered as a case of 'Exceptional circumstances' for reconstruction? In: Nagaoka M (ed) The future of the Bamiyan Buddha statues. Heritage reconstruction in theory and practice, pp 239–256. Springer-UNESCO, Paris

Praxenthaler B, Beckh M (2020) Safeguarding and preservation activities at the giant Buddhas and other monuments in the Bamiyan Valley 2004–2017. In: Nagaoka M (ed) The future of the Bamiyan Buddha statues. Heritage reconstruction in theory and practice, 31–50. Springer-UNESCO, Paris

Prott LV (2006) The protection of cultural movables from Afghanistan: developments in international management. In: Van Krieken-Peters J (ed) Art and archeology of Afghanistan. Its fall and survival, handbook of oriental studies, section eight: Central Asia, vol XIV, pp 189–200. Brill, Leiden-Boston

Ringbeck B (2022) World Heritage and Reconciliation. In: Marie-Theres Albert MT, Bernecker R, Cave C, Prodan AC, Ripp M (eds) 50 Years world heritage convention: shared responsibility-confict & reconciliation. Springer, Federal Foreign Office, Germany, pp 439–444

Rössler M (2020) World heritage and reconstruction: an overview and lessons learnt for the Bamiyan Valley. In: Nagaoka M (ed) The future of the Bamiyan Buddha statues. Heritage reconstruction in theory and practice, pp 99–112. Springer-UNESCO, Paris

Ruttig T (2011) The destruction of the Bamian Buddhas, vol 1. Afghan Analysts Network, https://www.afghanistan-analysts.org/en/reports/context-culture/the-destruction-of-the-bamian-buddhas-1/. Last Accessed 16 May 2016

Semple M (2011) Why the Buddhas of Bamian were destroyed. Afghan Analysts Network, https://www.afghanistan-analysts.org/en/reports/context-culture/guest-blog-why-the-buddhas-of-bamian-were-destroyed/. Last Accessed 19 Nov 2018

Sengupta R, Lal BB, Singh SB (2006) A report on the preservation of Buddhist monuments at Bamiyan in Afghanistan. Islamic Wonders Bureau, New Delhi

Toubekis G, Jansen M, Jarke M (2017) Long-term preservation of the physical remains of the destroyed Buddha figures in Bamiyan (Afghanistan) using virtual reality technologies for preparation and evaluation of restoration measures. ISRPS Ann Photogramm, Remote Sens Spat Inf Sci 4(2):271–27. https://www.researchgate.net/publication/319283287_LONG-TERM_PRESERVATION_OF_THE_PHYSICAL_REMAINS_OF_THE_DESTROYED_BUDDHA_FIGURES_IN_BAMIYAN_AFGHANISTAN_USING_VIRTUAL_REALITY_TECHNOLOGIES_FOR_PREPARATION_AND_EVALUATION_OF_RESTORATION_MEASURES. Last Accessed 11 Nov 2021

Toubekis G, Jansen M, Jarke M (2020) Physical revitalization of the Eastern Buddha statue in Bamiyan using reinforced adobe material. In: Nagaoka M (ed) The future of the Bamiyan Buddha statues. Heritage reconstruction in theory and practice, pp 307–330. Springer-UNESCO, Paris

UNESCO (1998) World Heritage Committee 21st session, Naples, Italy, 1–6 December 2017, report, WHC-97/CONF.208/17, Paris

Van Krieken-Peters J (2006) Dilemmas in the cultural heritage field: the Afghan case and the lessons for the future. In: Van Krieken-Peters J (ed) Art and archeology of Afghanistan. Its fall and survival, handbook of oriental studies, section eight: Central Asia, vol XIV, pp 201–227. Brill, Leiden-Boston

Van Linschoten AS, Kuehn F (2012) An enemy we created: the myth of the Taliban/Al Qaeda merger in Afghanistan 1970–2010. Hurst, London

Warnke M (ed) (1973) Bildersturm. Die Zerstörung des Kunstwerks, Hanser, München

The Archaeological Sites of Bamiyan Valley—Conditions and Possible Interventions from a Conservative Perspective

Fabio Colombo and Elisa Pannunzio

Abstract The following paper contains a detailed condition assessment of the main Bamiyan Valley's archaeological sites that have been visited from February 19th to 22nd, 2023, on behalf of UNESCO and of the Japan Archaeological Mission in Afghanistan. The survey involved all the main archaeological sites of the Bamiyan Valley: the main cliff with the giant Buddhas and many caves, the Kakrak and Folladi Valleys, Shahr-e Gholghola and Shahr-e Zohak. The purpose was to assess their current condition and to define priorities for immediate response actions, to prevent vandalic actions, the collapse of archaeological structures and further deterioration of the wall paintings. For each site, a detailed photographic documentation of the current condition was completed, and the main conservation issues have been identified. The information about each site is displayed in dedicated paragraphs, included in this document.

Keywords Bamiyan · Wall paintings · Archaeological structures

1 Overview and Method

The following paragraphs describe the condition assessment of the main archaeological sites of Bamiyan Valley, completed between February 19th and 22nd, 2023. The survey, planned by the UNESCO Office in Kabul and by the Japan Archaeological Mission in Afghanistan, comprehended:

- The two Giant Buddha niches and some of the adjacent caves in the Bamiyan Valley's Great Cliff.
- The Folladi Valley Caves.
- The Kakrak Valley Caves.

F. Colombo (✉)
Freelance Conservator, Via Delle Viole 14, 00172 Rome, Italy
e-mail: fabiocolombo69@gmail.com

E. Pannunzio
Freelance Conservator, Via Pietro Mascagni 27, 30171 Venice, Italy

© The Author(s) 2024
M. Loda and P. Abenante (eds.), *Cultural Heritage and Development in Fragile Contexts*, Research for Development, https://doi.org/10.1007/978-3-031-54816-1_4

51

– Shahr-e Gholghola.
– Shahr-e Zohak.

As is widely known, after August 2021 the maintenance and conservation programs in Bamiyan have stopped and the archaeological sites have been exposed to extreme weather conditions, to vandalic actions and, often, to a general lack of surveillance.

Aiming at the resume of conservation and maintenance activities, this survey served as a first general evaluation of the sites' priorities, to define possible strategies that could improve the condition of the most valuable and endangered structures.

In particular, the condition of the caves' wall paintings has already been monitored during past surveys and dedicated interventions, on behalf of UNESCO and of the Japan Archaeological Mission in Afghanistan. This last survey represents an update of the previous experiences on the Great Cliff's caves.[1]

For each site, a detailed photographic documentation was completed, and the main conservation issues have been identified. The condition of each site is described in dedicated paragraphs of this document together with proposals for immediate response actions.

The proposed strategies consider the involvement of local population, both to enhance the local economy and to allow for a continuous maintenance of the sites' areas, their doors, gates and draining systems.

2   Archaeological Sites Condition

2.1   Archaeological Site: East Giant Buddha Niche and Caves

Date of the survey. February the 20th, 2023.

Surveyed area. East Giant Buddha niche, its adjacent caves and the stairs and passages to them.

Access to the site. The site can be freely accessed as there are no guards and no gates. Not all the existing doors to the caves are locked.

Condition of the site and of its structures. The fragments of the giant sculpture at the bottom of the niche are still under wooden structures. However, fruit peels are visible on the structures' roofs and around them. This means that visitors freely enter the site, leaving waste and climbing on the covering structures without any surveillance.

---

[1] The most recent survey was completed in July 2017, and it consisted of the meticulous evaluation of the caves' wall paintings condition. It was completed by the conservators Fabio Colombo and Michela Gottardo.

The niche of the Giant Buddha suffered terrible damage in 2001. Currently, the fractures that cross the niche's vault do not show signs of recent water leaking from the cliff's wall. The gap filling and edging of the fractures are in good condition.

The structure of the internal stairs that leads to the top of the niche is in good condition.

The hooks used to climb the niche walls, to check the sculpture's remaining surface, are still pierced in the niche's stone. However, they require a careful inspection.

Condition of the wall plasters and paintings. The remnants of the giant sculpture show some areas where the finishing plasters of the sculpture are conserved. They have been consolidated in the past and their edgings are in good condition.

The absence of locked doors to the entrance of the stairs and of the caves' fore-courts is surely an issue for the conservation of the site. Traces of waste, graffiti and drawings drafted by scratching the ancient plasters are diffused on the caves' internal and external walls. For example, the presence of blue spray paint graffities are visible all along the entrance of the stairs and on the external and internal walls of many caves.

This is a clear sign of the lack of maintenance and surveillance of the site. Given the fragile and porous nature of the cliff's stone and of the plasters, graffiti removal may cause scratches and stains on the surfaces. Thus, it is important to avoid any further vandalic actions (Fig. 1).

Fig. 1 Forecourt's wall of Cave D1. Large blue spray paint graffities have been traced on this wall as well as on many other caves and forecourts' walls around the East Giant Buddha

Well-built doors can also prevent the entrance of small animals and birds inside the caves and thus avoid the deposit of animal droppings and guano on the decorated surfaces.

Recommendations for immediate response actions:

– Regular surveillance of the site, as visitors should not be allowed to enter the site without a guide.
– Construction of gates and fences around the base of the Buddha niche, to prevent visitors from entering without any surveillance, especially inside the caves and around the fragments of the Giant Buddha.
– Construction of new locked doors to close the entrances to the stairs, to the caves and their forecourts.
– Maintenance of the existing doors that, sometimes, are not locked or still allow small animals and birds to access the caves.
– Inspection and maintenance of the drainage system of the rainwater from the cliff.
– Plan tests for spray paint graffiti removal protocols.

2.2 Archaeological Site: Cave D1

Date of the survey. February the 20th, 2023.

Surveyed area. Cave D1, the lowest of the caves' D group, and its forecourt. The cave has decorated plasters while its forecourt has unpainted plasters.

Access to the cave. The cave has no door as well as its forecourt.

Condition of the site and of its structure. As the cave has no locked door, it can be freely accessed by visitors. Thus, waste and food peels are visible on the forecourt and cave's floors.

The forecourt and cave's walls present fractures that cross the cliff stone and the decorated plasters.

Between the forecourt and the cave's entrance, there are some emerging circular structures. As their function is not known, they should be carefully documented and studied by archaeologists. Once carefully documented, they require consolidation and reburial.

Condition of the wall plasters and paintings. Marker writings, incisions and graffities are visible on the surfaces of the cave and forecourt's plasters.

The unpainted plasters on the forecourt's walls have no edgings thus they may have not been consolidated in the past.

The cave's painted platers, instead, have been consolidated in the past as edgings are visible. However, small areas where the edgings are detaching from the stone walls or where the plasters lack any edging, possibly due to recent detachment, are visible.

Recommendations for immediate response actions:

- Regular surveillance of the site, as visitors should not be allowed to enter the site without a guide accompanying them.
- Construction of new locked doors to close the entrances to the cave and to its forecourt.
- Plan and complete tests of spray paint graffiti removal protocols.
- Set up of crack meters, to monitor the condition of the fractures of walls and plasters.

2.3 Archaeological Site: Cave D

Date of the survey. February the 20th, 2023.

Surveyed area. Cave D, the highest of the caves' D group, and its forecourt. Both decorated with painted plasters.

Access to the cave: The cave has no door as well as its forecourt.

Condition of the site and of its structure. The forecourt and cave's walls present fractures that cross the cliff stone and the decorated plasters (Fig. 2).

Condition of the wall plasters and paintings. As the cave has no locked door, it can be freely accessed by visitors. Thus, blue spray paint graffities are visible on the walls of the forecourt.

Fig. 2 Detail of Cave D frame, at the base of the dome. Fractures cross the stone walls and the painted plasters, causing their detachment and loss

The painted plasters on the forecourt's walls have been treated and consolidated in the past: their surface appears in good condition as well as their edging repairs.

The cave's painted platers have also been consolidated in the past as edgings are present. However, there are diffused areas where wall painting detachments are visible, revealing the cave's stone surface that has a lighter color, as it is not covered by dust deposits. The silhouette of the detached plasters is highlighted by their edging mortars, still adhering to the stone surface.

Leaking of guano is visible on the frame at the base of the dome, staining the wall plasters and stucco decorations.

Recommendations for immediate response actions:

– Regular surveillance of the site, as visitors should not be allowed to enter the site without a guide accompanying them.
– Construction of a new locked door to close the entrances to the cave and to its forecourt.
– Plan and complete tests of spray paint graffiti removal protocols.
– Set up of crack meters, to monitor the condition of the fractures of walls and plasters.

2.4   Archaeological Site: Cave Ca

Date of the survey. February the 20th, 2023.

Surveyed cave. Ca (165) and its forecourt, both decorated with painted plasters.

Access to the cave. The cave has a locked door while the forecourt can be freely accessed.

Condition of the site and of its structure. The door of the cave is in good condition and locked. It does not allow small animals and birds to enter the cave.

Condition of the wall plasters and paintings. As the forecourt has no locked door, it can be freely accessed by visitors. Thus, blue spray paint graffities, marker writings and incisions are visible and diffused on the walls and plasters of the forecourt.

The painted plasters have been treated and consolidated in the past: except for the graffiti, their surface appears in good condition as well as their edging repairs.

Leaking of guano is visible on the frame at the base of the dome, staining the wall plasters and stucco decorations.

Recommendations for immediate response actions:

– Regular surveillance of the site, as visitors should not be allowed to enter the site without a guide accompanying them.
– Construction of a new locked door to close the entrance to the forecourt.
– Plan and complete tests of spray paint graffiti removal.

- Completion of new edgings where recent detachments of the cave's plasters are visible.

2.5 Archaeological Site: Cave Cb

Date of the survey. February the 20th, 2023.

Surveyed area. Cave Cb (164) and its forecourt. Decorated plasters are present only on the cave's walls and ceiling.

Access to the cave. The forecourt has no door while the cave has a locked one.

Condition of the structure. Fractures are visible on the cave's walls. A sign in the cave informs about the presence of an environmental monitoring system.

Condition of the wall plasters and paintings. The wall paintings of the cave have been consolidated in the past, by the Archaeological Survey of India, between 1969 and 1978. They appear in acceptable condition, but they should receive maintenance and treatment as many years have passed since the last interventions.

The cave's door does not have a perfectly sealed frame thus small animals and birds can still enter the cave. Guano leaking is visible on the cave's frame.

Small white writings are visible on the decorated plaster of the cave, and they have been probably done with chalk.

On the floor of the cave, remnants of the ancient stucco floor have been identified. They should be consolidated and carefully documented as it is very rare to find large fragments of these floor finishing.

Recommendations for immediate response actions:

- Regular surveillance of the site, as visitors should not be allowed to enter the site without a guide accompanying them.
- Construction of a new locked door to close the entrance to the forecourt. Maintenance of the existing cave door, that has a not perfect frame sealing.
- Plan and complete tests of chalk writing removal protocols.
- Set up of crack meters, to monitor the condition of the fractures of walls and plasters.
- The environmental monitoring system should be resumed, and its data analyzed.

2.6 Archaeological Site: Group B Caves

Date of the survey. February the 20th, 2023.

Surveyed area. Group B (138–139) caves and forecourts. They all have decorated plasters.

Access to the caves. Only cave Ba has a locked door.

Condition of the structure and of the site. Fractures are visible on walls and ceilings of the forecourt and of the caves.

Condition of the wall plasters and paintings. The forecourt of the caves and cave n. 139 can be freely accessed, thus spray paint graffities, marker writings, scratches and other vandalic actions are visible and diffused on the forecourts' walls and plasters.

The plasters have edgings. However, they should be carefully inspected to verify the presence of recent detachment.

Recommendations for immediate response actions:

- Regular surveillance of the site, as visitors should not be allowed to enter the cave and balcony without a guide accompanying them.
- Construction of a new locked door to close the entrance to the forecourts and to cave 139. Maintenance of the existing cave door, that has a not perfect frame sealing.
- Plan and complete tests of graffities removal protocols.
- Set up of crack meters, to monitor the condition of the fractures of walls and plasters.

2.7 Archaeological Site: Caves Group A Upper

Date of the survey. February the 20th, 2023.

Surveyed area. Caves group A Upper (Aa 129, 130), caves and forecourts.

Access to the caves. Cave Aa (129) has a locked door.

Condition of the structure and of the site. Fractures are visible on walls and ceilings of the forecourts and of the caves.

Condition of the wall plasters and paintings. Even though Cave Aa (129) has a locked door, it is sadly known for the vandalic action it has been subjected to: the surface of the decorated plaster, darkened by the dust deposit, is covered with soles prints.

Cave 130 has not a door. The surface of its plasters, even though covered by a layer of whitewash, present a huge number of spray paint graffiti, marker and chalk writings and scratches.

Recommendations for immediate response actions:

- Regular surveillance of the site, as visitors should not be allowed to enter the cave and balcony without a guide accompanying them.
- Construction of new locked doors to close the entrance to the forecourts and to cave 130. Maintenance of the existing door, that has a not perfect frame sealing.
- Plan and complete tests of graffiti removal protocols.
- Set up of crack meters, to monitor the condition of the fractures of walls and plasters.

2.8 Archaeological Site: West Giant Buddha Niche

Date of the survey. February the 20th–21st, 2023.

Surveyed area. West Giant Buddha niche.

Access to the site. The site can be freely accessed as there are no guards and no functioning fences and gates. Not all the existing doors to the caves are locked.

Condition of the site and of its structure. The fragments of the giant sculpture at the bottom of the niche are still under the wooden shelters that host them. However, it is visible how the structures and materials stocked beneath them have been voluntarily damaged and looted. Moreover, the shelters were already unsuitable for the safe storage of the fragments, as they are not isolated from the environment, and they cannot prevent future deterioration of the archaeological materials.

Many of the caves and the forecourts have no locked doors. Well-built doors can avoid further vandalic actions as well as the entrance of small animals and birds inside the caves and the deposit of animal droppings and guano on the decorated surfaces.

The niche of the Giant Buddha suffered terrible damage in 2001. Currently, the fractures that cross the niche's vault do not show signs of recent water leaking from the cliff's wall. However, a careful monitoring of its fractures is recommended as well as the inspection of the scaffoldings and security nets that have been installed as temporary supporting structures for the remnants of the giant sculpture.

Recommendations for immediate response actions:

– Regular surveillance of the site, as visitors should not be allowed to enter the site without a guide accompanying them.
– Construction of gates and fences around the base of the Buddha, to prevent visitors from entering without any surveillance, especially inside the caves and around the fragments of the Giant Buddha.
– Construction of new locked doors to close the entrances to the stairs, to the caves and their forecourts.
– Maintenance of the existing doors that, sometimes, are not locked or still allow small animals and birds to access the caves.
– Inspection and maintenance of the drainage system of the rainwater from the cliff.
– Plan and construction of more suitable shelters for the Giant Buddha fragments.

2.9 Archaeological Site: Cave at the Base of the West Giant Buddha Niche

Date of the survey. February the 20th, 2023.

Surveyed area. Cave at the base of the West Buddha, Nort East corner. It hosts fragments of the giant Buddha and has decorated plasters on the walls and ceiling.

Fig. 3 Cave at the base of the West Buddha, Nort East corner. It hosts fragments of the giant Buddha and has decorated plasters on the walls and ceiling. The cave has been vandalized: the giant sculpture's fragments have been removed from their wooden supports and damaged

Access to the cave. The cave has no door and can be freely accessed.

Condition of the structure and of the site. The cave has been vandalized. The fragments of the Buddha sculptures have been damaged, removed from their wooden supports and possibly looted (Fig. 3).

Condition of the wall plasters and paintings. The wall plasters have been consolidated in the past as edgings are visible. However, given the absence of a door and the vandalic actions perpetrated inside the cave, their conditions should be carefully inspected.

Recommendations for immediate response actions:

– Regular surveillance of the site, as visitors should not be allowed to enter the site without a guide accompanying them.
– Construction of a new locked door to close the entrance to the cave.
– Plan and complete the construction of more suitable shelters for the Giant Buddha fragments, to store them in upgraded conditions.
– Careful inspection of the wall plaster's condition.

2.9.1   Archaeological Site: Cave 53-V

Date of the survey. February the 20th, 2023.

Surveyed area. Cave 53-V (621) under the West Buddha niche.

Access to the cave. The cave has a locked door.

Condition of the structure and of the site. The door requires maintenance as small animals and birds can still enter the internal space.

Fractures are visible across the cave's walls.

The cave has been vandalized. The archaeological fragments stored inside the cave, piled on improper supports, have been damaged, removed from their supports and possibly looted.

Condition of the wall plasters and paintings. The wall plasters have been consolidated in the past as edgings are visible. However, given the vandalic actions perpetrated inside the cave, their conditions should be carefully verified.

Recommendations for immediate response actions:

- Regular surveillance of the site, as visitors should not be allowed to enter the site without a guide accompanying them.
- Maintenance of the door to prevent animals and birds from entering the cave.
- Plan and construction of more suitable shelters for the archaeological fragments, to store them in upgraded conditions.
- Set up of crack meters, to monitor the condition of the fractures of walls and plasters.
- Careful inspection of the wall plaster's condition.

2.9.2   Archaeological Site: Cave I

Date of the survey. February the 21st, 2023.

Cave. Cave I and its forecourt. They both have decorated painted surfaces.

Access to the cave. To access the forecourt it is necessary to transport and install a ladder. The cave's access has a locked door.

Condition of the structure and of the site. The water drainage system is still functioning. However, it requires maintenance and cleaning.

The forecourt has some carved, concave spaces where nets have been positioned to avoid birds' nests and guano deposits. Some of these nets are fallen on the floor and recent guano leaks are visible on the balcony's walls.

Fractures are visible on the forecourt walls.

Condition of the wall plasters and paintings. The wall plasters of the forecourt have been consolidated in the past as edging is visible and they are in good condition.

However, some water can still leak on the balcony's ceiling surface, and it creates soil encrustation and erosion.

Moreover, as the forecourt is exposed to the environment, the surface of the wall paintings is eroded and abraded.

The cave's wall paintings have been already consolidated too as edgings are visible and they are in good condition. However, numerous marker writings and scratches are visible on the decorated surfaces.

Recommendations for immediate response actions:

– Regular surveillance of the site, as visitors should not be allowed to enter the site without a guide accompanying them.
– Maintenance of the door and of the nets, to prevent animals and birds from entering the cave.
– Maintenance of the existing drainage system.
– The base of the removable ladder to access the balcony should be repaired as it is unstable.
– Set up of crack meters, to monitor the condition of the fractures of walls and plasters.
– To access and monitor the wall paintings, a system of pegs and supports should be installed in the cave and its balcony so that, whenever it is necessary, a scaffolding can be quickly built.

2.9.3   Archaeological Site: Caves N

Date of the survey. February the 21st, 2023.

Surveyed area. Caves N and their forecourts. They both have decorated painted surfaces.

Access to the cave. To access the forecourts it is necessary to transport and install a ladder. The cave's access has a locked door.

Condition of the structure and of the site. The water drainage system is still functioning, although it requires cleaning and maintenance.

The main forecourt does not have a locked door while the caves of this group all have functioning doors.

Waste and abundant guano deposits are visible on the forecourt floor and walls, and this means that visitors and animals freely enter this area.

Fractures are visible on the ceiling of the balcony.

Condition of the wall plasters and paintings. On the forecourts walls and ceiling small areas where plasters are still conserved are visible. However, these plasters do not have edgings and they immediately require consolidation treatments.

The caves are decorated with remarkably important paintings. It was preferred to avoid opening the locked doors to inspect them, both to prevent any change in the internal environment of the cave and to avoid the entrance of too many people.

Recommendations for immediate response actions:

– Regular surveillance of the site, as visitors should not be allowed to enter the site without a guide accompanying them.

- Set up of nets, to prevent animals and birds from entering the balcony.
- Maintenance of the existing drainage system.
- The base of the removable ladder to access the forecourt should be repaired as it is unstable.
- Set up of crack meters, to monitor the condition of the fractures of walls and plasters.
- Treatment of the plasters that are still present on the balcony walls.

2.9.4 Archaeological Site: Folladi Valley

Date of the survey. February the 21st, 2023.

Surveyed area. Folladi Valley and its caves.

Access to the caves. The caves have no doors as well as the site, can be freely accessed by visitors and animals.

Condition of the structure and of the site. The site is completely abandoned: there is no surveillance even if it is located near to a small village. Animals and local inhabitants use the caves as shelters so that graffiti and writings are visible on the walls and thick deposits of animal droppings cover the cave floors (Fig. 4).

Fractures are visible on caves walls and ceilings.

Fig. 4 One of the lower caves in Folladi Valley, used as dung storage

Condition of the wall plasters and paintings. The paintings are completely exposed to the environment, to erosion due to anthropic activities and to vandalic actions.

Recommendations for immediate response actions:

– Regular surveillance of the site, possibly involving the local community, as visitors should not be allowed to enter the site without a guide accompanying them.
– Set up of locked doors or nets to avoid the entrance of animals and visitors to the caves.
– Removal of animal dropping deposits from the caves.
– Edging repairs of the decorated plasters.
– Detailed documentation of the caves and of their surviving decorations as they are exposed to extreme environmental conditions (detailed photographic documentation, lasers scan, drone images, 3D models…).

2.9.5   Archaeological Site: Kakrak Valley

Date of the survey. February the 22nd, 2023.

Surveyed area. Kakrak Valley and its caves.

Access to the caves. The caves have no doors as well as the site, that can be freely accessed by visitors and animals.

Condition of the structure and of the site. The site is completely abandoned: there is no surveillance even if it is surrounded by cultivated fields. Local inhabitants use the caves as shelters.

Fractures are visible on cave walls and ceilings.

Some structures, such as a tower, are still standing even though they require immediate structural consolidation.

Condition of the wall plasters and paintings. The plasters are completely exposed to the environment and to erosion due to anthropic activities.

Recommendations for immediate response actions:

– Regular surveillance of the site, as visitors should not be allowed to enter the site without a guide accompanying them.
– Identify the structures more informative rich or important from an archaeological and artistical point of view. Set up of locked doors or fences to prevent the entrance of animals and visitors to these identified areas.
– Removal of animal dropping deposits from the caves.
– Edging repairs of the plasters.
– Detailed documentation of the caves and of their surviving decorations as they are exposed to extreme environmental conditions (detailed photographic documentation, laser scans, drone images, 3D models…).

2.9.6   Archaeological Site: Shahr-e Gholghola

Date of the survey. February the 19th, 2023.

Surveyed area. SW and SE slopes of the hill, with a particular focus on the areas 2, 3, 22, 19, 15 and on the Citadel.[2]

The site has been documented entering from its SW side, the current entrance, where pictures have been taken of the high and complex structures still standing along the current entrance. Then, the survey proceeded backward and then upward, following the current path up to the so called "Citadel", the highest cluster of structures still standing on the top of the hill.

Once documented this part, the survey proceeded down along the E wall of the Citadel and down again, along the N side of the cliff, where many structures are still standing along the huge empty space left by the erosion of the N side of the cliff.

Then, another area that has been carefully documented is the path that runs South and West ward, towards the ICOMOS restored area, where are located many high structures, an arch with colored plasters and a series of low caves, with traces of overlapped painted layers and plasters.

Access to the site. The site can be freely accessed as there are no guards and no gates.

Condition of the site and of its structures. The survey did not reveal dramatic changes in the structures' condition nor major collapses happened since the last extensive survey of the site, completed in October 2019.

However, the structures show evident signs of erosions due to the extreme environmental conditions they are exposed to, and many buildings are collapsing.

The site lacks a regular maintenance and an extensive and functional drainage system. Thus, rain and melting snow erode the lower portion of the buildings, weakening their structures and increasing the potential of their collapse (Fig. 5).

Condition of the wall plasters and paintings. The plasters of the structures represent an important source of archaeological information about the site's history as they allow to spot overlapping structures and traces of no longer existing ceilings.

Again, the survey did not reveal extensive loss of plasters. However, they have never been consolidated so far nor have edgings to protect them from water leaking and detachment. Thus, an extensive intervention of edging and consolidation is necessary, given the erosion's speed of the site.

Recommendations for immediate response actions:

– Regular surveillance of the site, as visitors should not be allowed to enter the site without a guide accompanying them.

---

[2] The site anagraphica are the ones defined during the 2019 extensive survey of the site, "Assessment, design, and supervision for the conservation of archaeological remains at Shahr-e Gholgola", completed by the Shahr-e Gholghola Italian Consultancy team for UNESCO (contract for services no. 4500405580-A2).

Fig. 5 Shahr-e Gholghola, SW slope of the hill. Some high structures are still standing, partially covered by plasters. There is a high risk of collapse due to erosion caused by environmental factors and to the steepness of the hill

– Set up of fences or nets and gates around the hill, to avoid uncontrolled entrances to the site.
– Plan and construction of the site's drainage system.

2.9.7 Archaeological Site: Shahr-e Zohak

Date of the survey. February the 22nd, 2023.

Surveyed area. During the survey, it was possible to visit only the lower citadel, on the East and West escarpments. The steep paths to the higher citadel were not safe because of the ice on the ground.

Access to the site. The site can be freely accessed as there are no guards and no gates.

Condition of the site and of its structures. The structures show evident signs of erosions due to the extreme environmental conditions they are exposed to, and many buildings are collapsing. The site lacks a regular maintenance and an extensive and functional drainage system. Thus, rain and melting snow erode the lower portion of the buildings, weakening their structures and increasing the potential of their collapse (Fig. 6).

Fig. 6 Shahr-e Zohak lower citadel and, on the background, part of the structures of the higher citadel

Condition of the wall plasters and paintings. The plasters of the structures represent an important source of archaeological information about the site's history as they allow to spot overlapping structures and traces of no longer existing ceilings.

The plasters and the architectural surfaces have never been consolidated so far nor have edgings to protect them from water leaking and detachment. Thus, an extensive intervention of edging and consolidation is necessary, given the erosion's speed of the site.

Recommendations for immediate response actions:

– Regular surveillance of the site, as visitors should not be allowed to enter the site without a guide accompanying them.
– Detailed documentation of the site and of the archaeological structures as they are exposed to extreme environmental conditions and the risk of their collapse is high (detailed photographic documentation, laser scan, drone images, 3D models…).
– Completion of fences and gates around the hill, to avoid uncontrolled entrances to the site.
– Planning and completion of the site's drainage system.

3   Condition of the Sites and Main Conservation Issues

The survey allowed to identify common issues and threats to the conservation of the surveyed Bamiyan sites, of the surviving structures, plasters, wall paintings and decorative elements:

– Lack of surveillance. Many sites can be freely accessed such as Shahr-e Zohak and the caves of the minor valleys, Kakrak and Folladi. Even though the Great Cliff and Shahr-e Gholghola have guards at their entrances, they do not stop visitors from freely entering the sites, from climbing the archaeological structures and, sometimes, perpetuating vandalic actions (graffiti).
– Vandalic actions. The storages and caves of the West Buddha, that contain fragments of the giant sculpture, have been vandalized and looted. Unfortunately, also the East Giant Buddha caves have been heavily targeted with vandalic actions (graffiti, scratches and small damages), especially the caves' forecourts, more visible and accessible.
– Absence of locked doors. The entrance doors to the East Giant Buddha caves are open, as well as to the Folladi caves, that are currently used as storages and animals shelters (manure covers the cave's floor). Securing the access to the caves with locked doors represents an efficient solution to avoid intentional damages or incidents which could in any way endanger the sites. The caves that, in the past, have been secured with locked doors have not been damaged recently.
– Lack of maintenance. The sites are often dirty with paper remains, fruit peels, bottles, food remains, cigarette wraps and butts.
– Erosion, due to the extreme environmental conditions, is especially visible in Shahr-Gholghola and Shahr-e Zohak. However, damages due to erosion are dramatically visible on all the archaeological sites visited during the survey, where there is a general deterioration of the condition of both the decorative and structural elements.

4 Proposals for a Maintenance Plan and for Immediate Response Actions

The condition of the sites, as resulted during the survey, suggests two necessary strategies to prevent further damage to the archaeological heritage of the area.

First of all, there are immediate response actions, that consist of targeted interventions aiming at the increase of security and surveillance of the sites, as well as to mitigate the deterioration caused by anthropic actions and by the extreme weather conditions of the region:

– To secure all existing doors of caves and sites as well as to build and set up doors for the caves that are not closed yet.
– To prevent indiscriminate and uncontrolled access, delimiting the most important archaeological areas with gates and nets.
– To train and hire local staff to provide effective surveillance.
– To resume ordinary and extraordinary maintenance of the entire water drainage system of the Great Cliff and to implement the water drainage systems to the sites that currently lack any of these systems (Shahr-e Gholghola, Shar-e Zohak, Kakrak and Folladi valleys).
– To plan graffiti removal operations, to avoid permanent spots and damages to the wall paintings.
– To plan and complete targeted consolidation intervention of wall paintings and plasters (edging, grouting and eventually temporary facing) and structural consolidation or temporary support for collapsing archaeological architectures.
– To plan the complete documentation of collapsing and endangered archaeological structures in the Kakrak and Folladi Valleys as well as in Shahr-e Gholghola and Shahr-e Zohak. This is to prevent the loss of any information, in case any damage occurs to the sites.

These operations can be planned in accordance with the priorities identified during the survey.

On the other hand, it is necessary to define a plan for the general maintenance of the sites:

– Skilled workers can be hired to provide for the maintenance of the doors and general cleaning of the sites.
– Local guides should be trained and hired to accompany visitors.
– The involvement of local community in the correct management of the sites and of the buffer zones should be encouraged. This can provide both improved maintenance of the sites as well as an economic source of income for the local population.
– Monitoring of the urban and rural expansion and the buffer zones to the sites.

Maintenance operations also comprehend the resume of monitoring programs. Some of the caves already have environment monitoring systems that should be checked and resumed. Cracks and fractures visible on the Cliff and on the caves' walls should be monitored too.

Finally, it would be beneficial to resume, also, the conservation, research and archaeology programs. This, aims at a deeper knowledge of the sites and at better conservation conditions of Bamiyan archaeological heritage, enhancing its possibility to be preserved throughout this unpredictable situation.

Requirements for the Protection of the UNESCO World Heritage Cultural Landscape and Archaeological Remains of the Bamiyan Valley (Afghanistan)

Georgios Toubekis⦿

Abstract The international efforts to establish a management system for the UNESCO World Heritage Bamiyan in Afghanistan, where fundamentalist terrorists once destroyed the world's largest Buddhist figures, are discussed. A Cultural Master Plan based on remote sensing methods is being used to create a comprehensive inventory of archaeological zones, traditional settlements, water systems, and historical cultural monuments, and to monitor the effectiveness of conservation measures and urban development decisions on landscape change. The case of Bamiyan exemplifies the increasing politicization of outstanding heritage sites around the world that are subject to ongoing conflict, targeted iconoclastic raids, and criminal looting, among dynamic urban development and limited resources for conservation, destroying such sites and landscapes. Lessons learned highlight the importance of careful planning and management of development activities as an integral part of heritage conservation in international relief efforts in crisis areas, especially for World Heritage cultural landscapes. In the context of reconstruction efforts, the management of cultural heritage needs to consider a broader context that considers the identity of the people, the participation, and the revitalization of the affected communities in addition to the long-term preservation of the physical substance. Such an expanded notion of heritage recognizes local communities' aspirations for inclusive social and economic development considering Landscape Values as Human Values for reaching Sustainable Development Goals (SDG).

Keywords Afghanistan · UNESCO World Heritage · Remote sensing · Cultural heritage management · Reconstruction · Sustainable development

G. Toubekis (✉)
Fraunhofer Institute for Applied Information Technology FIT, RWTH Aachen University, Ahornstr. 55, 52074 Aachen, Germany
e-mail: georgios.toubekis@fit.fraunhofer.de

© The Author(s) 2024
M. Loda and P. Abenante (eds.), *Cultural Heritage and Development in Fragile Contexts*, Research for Development, https://doi.org/10.1007/978-3-031-54816-1_5

Fig. 1 Topographical map of the Bamiyan World Heritage nomination—eight distinct property areas (red line) each with a buffer zone (blue line) (*Source* UNESCO)

1 Introduction

The Cultural Landscape and Archaeological Remains of the Bamiyan Valley were nominated a UNESCO World Cultural Heritage in 2003 according to the 1972 World Heritage Convention (UNESCO 1972) and inscribed on the List of World Heritage in Danger, comprising a serial nomination of a total of eight separate areas[1] (Fig. 1).

Since 2004, UNESCO has been active in preserving this World Heritage Site through an internationally funded safeguarding campaign (UNESCO World Heritage Centre 2012) addressing urgent conservation needs. Achievements included the stabilization of the fragile cliffs and niches after the removal of the Buddha figures. The landscape management planning tools were developed at a time when the country had a limited number of legal instruments that could withstand the challenges of an ever-changing political environment. Despite the evolving circumstances, these tools have consistently proven their robustness over time and are now in need of revision to support the creation of a landscape management system necessary for the long-term preservation of the cultural landscape and its archaeological remains.

This introduction addresses the global challenges of heritage management and sets the context for heritage conservation in Afghanistan, specifically Bamiyan. The next section discusses the significance and protection of the Bamiyan World Heritage Site under the UNESCO World Heritage Convention. The third section describes the Cultural Master Plan for Bamiyan, detailing its objectives, methodology, and protective zoning proposal. The article concludes by drawing lessons from Bamiyan and offering a specific European perspective for future international assistance to Afghanistan's cultural heritage.

---

[1] See World Heritage Information System: https://whc.unesco.org/en/list/208/documents/.

1.1 Global Challenges to Cultural Heritage Management

Cultural heritage preservation in the 21st century faces a multitude of challenges, causing detrimental impacts on heritage sites and calling for the implementation of adaptive measures integrated within overall development strategies.

Environmental factors, particularly global climate change, with its changing weather patterns, have led to a rise in the frequency and intensity of natural disasters, endangering the structural integrity of monuments and archaeological remains (UNESCO World Heritage Centre 2014; Sabbioni et al. 2010).

Human factors, alongside environmental ones, such as vandalism and intentional destruction undermine the integrity and authenticity of heritage sites. Illegal excavations and the illicit trade of stolen antiquities exacerbate this issue, contributing to organized crime and illegal activities. The 1970 *UNESCO Convention on the Means of Prohibiting and Preventing the Illicit Import, Export, and Transfer of Ownership of Cultural Property* addresses these challenges through international cooperation and awareness campaigns for stricter regulations. Encouraging more countries to ratify and implement the convention is crucial for enhancing the impact of these efforts in combating the illicit trade of cultural property.[2]

Combat impunity for heritage-related war crimes, acknowledges the legal connection between the protection of cultural heritage and the prosecution of human rights abuses, as highlighted by examples such as Bamiyan, Palmyra, and Timbuktu. Preserving and providing access to cultural heritage are inherent rights for individuals and communities, demanding appropriate legal frameworks. International conventions like the Hague and UNESCO's World Heritage Convention aim to safeguard cultural heritage during conflicts, but enforcement remains a challenge (Francioni and Gordley 2013; Silverman and Ruggles 2007).

Adequate resource allocation for cultural heritage, to enhance regular maintenance, as aging and inadequate preservation are pressing concerns. Professional training, skills development, and capacity building are urgently needed to balance conservation, accessibility, and sustainable development needs. Limited resources pose a substantial impediment to achieving comprehensive efforts and attaining long-term results, as evidenced by numerous global examples (Bumbaru et al. 2000; Machat et al. 2020).

1.2 Balancing Heritage Management and Development in Bamiyan

Afghanistan, a diverse multi-ethnic, multireligious, and multicultural society, has experienced internal decade-long conflicts leading to global repercussions. The

---

[2] The current status of its implementation can be overlooked here: https://www.unesco.org/en/legal-affairs/convention-means-prohibiting-and-preventing-illicit-import-export-and-transfer-ownership-cultural last accessed 2023/05/28.

destruction of the Buddhas in 2001 despite a worldwide outcry (Manhart 2001) marked a turning point. After the fall of the previous Taliban regime, international support was promised, but the return of the Taliban in 2021 led to significant changes. Donors suspended projects due to concerns over the actions of the new Taliban powerholders, and the United Nations Assistance Mission in Afghanistan (UNAMA) expressed alarm over human rights violations (United Nations Assistance Mission in Afghanistan (UNAMA) 2022). This raises questions about future humanitarian aid and cultural heritage preservation. Addressing this requires a flexible strategy within the socio-political landscape, safeguarding both tangible and intangible cultural elements vital to Afghanistan's identity.

Reconstruction of destroyed Bamiyan Buddha figures. Reconstructing the Bamiyan Buddha figures has been a contentious issue for over two decades. Some suggest leaving the site as a war memorial, while others advocate reconstruction. This debate, along with lessons from other intentionally destroyed sites, has sparked international discussions. ICOMOS conducted studies on reconstructing World Heritage sites, emphasizing that it goes beyond physical restoration and can aid in trauma recovery, community revitalization, and heritage values appropriation. Recognizing it as a means of recovery, community involvement, and peacebuilding, these approaches aim to create a holistic framework that goes beyond physical restoration (ICOMOS 2017).

In Bamiyan, the previous Afghan government and local population long supported reconstructing at least one figure as a symbolic act of resurrection to create a major monumental attraction that also fits into the planned tourism strategy for the valley. Critical for long-term preservation is the recognition that the Bamiyan World Heritage encompasses an extensive cultural landscape, extending beyond the confines of the area of the destroyed Buddha figures. The entire overview of the debate along with contributions from international experts on reconstruction options is given in Nagaoka (2020).

New uncertainties. The valley's designation as a UNESCO World Heritage has aided overall state-building efforts and influenced identity-building processes, particularly for a younger generation now deprived of civilizational achievements after the collapse of the democratically elected government in August 2021. The recent political shifts have raised concerns, including reports from social media of direct threats to the site, such as RPG rocket firings towards the empty Buddha niches, vandalism, uncontrolled excavations, and commercial activities near the old bazaar.[3]

There are also credible reports of defamatory threats against local individuals previously supporting the safeguarding of the site. Given the complexity of working with de facto power holders, it is critical to consider how future international assistance should be structured to effectively address the current challenges of preserving cultural heritage not only in the Bamiyan Valley but throughout Afghanistan.

---

[3] See Decision: 45 COM 7A.49 of the Word Heritage Committee.

2 Protecting the Cultural Landscape of Bamiyan

2.1 Significance and Value of the Bamiyan World Heritage Property

The Bamiyan Valley is a remarkable fusion of artistic and religious developments that shaped the Central Asian region from the 1st to the 13th century CE. Bamiyan is considered a crossroads between the civilizations of the East and the West and was a historical center of early cross-cultural exchange between China and India. The valley served as a prominent center of Buddhist art along the Silk Road. The cave complexes and their iconic Buddha figures were symbols of cultural exchange and religious tolerance. The giant Buddha figures of Bamiyan, with their heights of 55 and 33 m, were considered the world's largest representation of standing Buddha figures.

The local population has not attached any religious significance to the figures for generations, as the Buddhist religion has not played any role in the region with the spread of Islam since the 11th century CE onwards. The figures were no longer considered Buddha depictions but have been reinterpreted and given new identities that have been integrated into the mythical story of the Islamization of Bamiyan (Klimburg-Salter 2020). The cultural landscape and archaeological remains represent a unique testimony not only of exceptional Buddhist art (6–9th centuries CE) but also of urban archaeological remains from the Islamic Ghaznavid and Ghurid periods (10–13th centuries CE).

Throughout history, Bamiyan has been the target of violent actions aimed at destroying iconic art and architecture. The cultural landscape of the Bamiyan Valley (about 2500 m above sea level) is surrounded by the imposing mountains of the Hindu Kush. It is characterized by a complex irrigation system based on a network of open earthen channels for the distribution of surface water. Outside the cultivated areas, steep mountains form a unique backdrop of a wild natural landscape. This arid land of gentle slopes consists of conglomerates and glacial sedimentary deposits that line the valleys engulfing steep cliffs. These barren lands with sparse vegetation are occasionally used for grazing or winter fuel. In stark contrast are the irrigated and intensively cultivated areas in the valley bottoms, where potatoes and robust cereals are grown. The visible archaeological remains complement this cultural landscape and have undergone few changes until recently (Fig. 2).

2.2 Insights from the UNESCO Safeguarding Campaign

The 2003 World Heritage nomination was a technical amendment to the earlier 1982 one, which solely focused on the archaeological features of the sparsely populated Bamiyan Valley. Although the World Heritage value was recognized, a lack of

Fig. 2 The Bamiyan Valley's Cultural Landscape in 2015, with its major cliff and the empty niche in the distance once occupied by the Giant Buddha figure (*Source* G. Toubekis)

management plans and updated reports delayed its nomination as World Heritage.[4] In 2003, after the Buddhas' destruction, Bamiyan was inscribed as a serial nomination with eight distinct areas as a cultural landscape,[5] however lacking documentation on its defined quality and precise extent. The UNESCO Safeguarding Campaign launched in 2003 aimed to support conservation and long-term preservation. ICOMOS Germany and RWTH Aachen University actively participated from its inception, extending efforts beyond the Buddhas' remains to establish an effective management system in line with the Operational Guidelines[6] of the World Heritage Convention. The goal was to remove the property from the endangered list within a decade, requiring tailored, site-specific approaches for long-term preservation and conservation.

Desired State of Conservation. The campaign focused on preserving Bamiyan's tangible cultural heritage through meticulous restoration and maintenance efforts, including safeguarding the remaining Buddha figure fragments, sculptured original surfaces, and cave paintings (Petzet 2009; Emmerling and Petzet 2016). The desired state of conservation adopted by the World Heritage Committee[7] for removing the property from the List of World Heritage in Danger is defined as the following:

1. ensured site security,

---

[4] See UNESCO World Heritage Committee Decision 19 BUR VI.24 (1995).

[5] The recognition of cultural landscapes in the framework of the 1972 UNESCO Convention for Safeguarding of the World Cultural and Natural Heritage was established in 1992.

[6] The Operational Guidelines for the Implementation of the World Heritage Convention are periodically revised to reflect the decisions of the World Heritage Committee and are accessible at: https://whc.unesco.org/en/guidelines/.

[7] Decision 31 COM 7A.21 from the 31st session of the World Heritage Committee (Christchurch, New Zealand, 2007).

Fig. 3 Various attributes of the Cultural Landscape of the Bamiyan Valley—religious shrines, vernacular earthen architecture, patterns of fields, and irrigation channels (*Source* RWTHacdc)

2. ensured the long-term stability of the Giant Buddha niches,
3. adequate state of conservation of archaeological remains and mural paintings,
4. implemented Management Plan and Cultural Master Plan (protective zoning plan).

Outstanding Universal Value (OUV). The cultural landscape is intricately connected to the harmonious blend of natural and human-made elements developed over centuries "as combined works of nature and of man".[8] This cultural imprint represents an interdependent system of cultural expression and human–environment interactions, carrying deep cultural and historical significance. It is crucial to consider conserving this cultural landscape amidst ongoing urban developments (Fig. 3).

Change Management. The cultural landscape evolves through "consistent, systematic and orderly decisions" (Rapoport 1992) of many individuals over time, rather than deliberate design. In an increasingly urbanized environment, how can these internal dynamics be preserved without freezing their current appearance? This landscape is vital for shaping identities, fostering cohesion, and ensuring cultural continuity for specific groups over time. Authenticity and integrity remain key considerations in its ongoing preservation.

3   The Cultural Master Plan for Bamiyan—Objectives and Strategies

In 2004, RWTH Aachen University collaborated with the Afghan government to create a plan identifying cultural heritage zones for protection within the context of urban development dynamics, to fulfill UNESCO preservation obligations. These protection zones were envisioned as distinct and noteworthy elements of the cultural landscape, essential for sustainably guiding future planning actions to avoid negative impact on the World Heritage property. The resulting *Bamiyan Cultural Master Plan* (*CMP*) addresses development challenges while emphasizing cultural and environmental sensitivity. Its main objective is to provide a planning framework that guides future projects to respect the Valley's cultural significance, ensuring its preservation as a UNESCO World Heritage Site (Jansen and Toubekis 2020). This protective

---

[8] Article 1 of the UNESCO 1972 World Heritage Convention.

zoning scheme was adopted by the Ministry of Urban Development jointly with the Ministry of Culture in 2006 and became part of the efforts to remove the property from the List of World Heritage in Danger (Jansen 2009).

3.1   Land Use Analysis and Inventory of the Cultural Landscape

The methodology involved national and international experts in assessing land uses, including agriculture, settlements, and infrastructure. Students at the local university were involved in studying the geomorphology of the valley. A thorough inventory of settlements, water systems, and cultural artifacts served as the foundation for the cultural master plan. The research resulted in a comprehensive plan identifying areas needing cultural protection, urbanized zones impacting the landscape, and potential areas for future urban development, considering the following three topics.

Authenticity. Through a combination of remote sensing analysis and extensive field surveys, key elements, and attributes of the cultural landscape, including historic areas, vernacular architecture, traditional settlements, and water canals, were documented. For the first time, this cultural inventory provided credible and sufficient information to meet the condition of authenticity required for World Heritage.[9]

Integrity. The property should also include a substantial portion of the elements that convey its full value to maintain relationships and functions within the cultural landscape that define its distinct character.[10] Continuous maintenance and effective management of the property's physical structure and notable attributes are essential while comprehending its full extent.

World Heritage Boundary. The analysis of the *Cultural Master Plan* has revealed the need to adjust the delineated areas of the World Heritage property beyond the initial nomination dossier. This boundary adjustment is crucial to ensure a sufficient portion of this significant cultural landscape is protected under the World Heritage Convention to consistently meet the long-term integrity criteria (Jansen and Toubekis 2007).

---

[9] The condition of authenticity is essential to convey the Outstanding Universal Value of a World Heritage property—see paragraphs 79–86 of the Operational Guidelines.

[10] See paragraphs 87–95 of the Operational Guidelines to the World Heritage Convention.

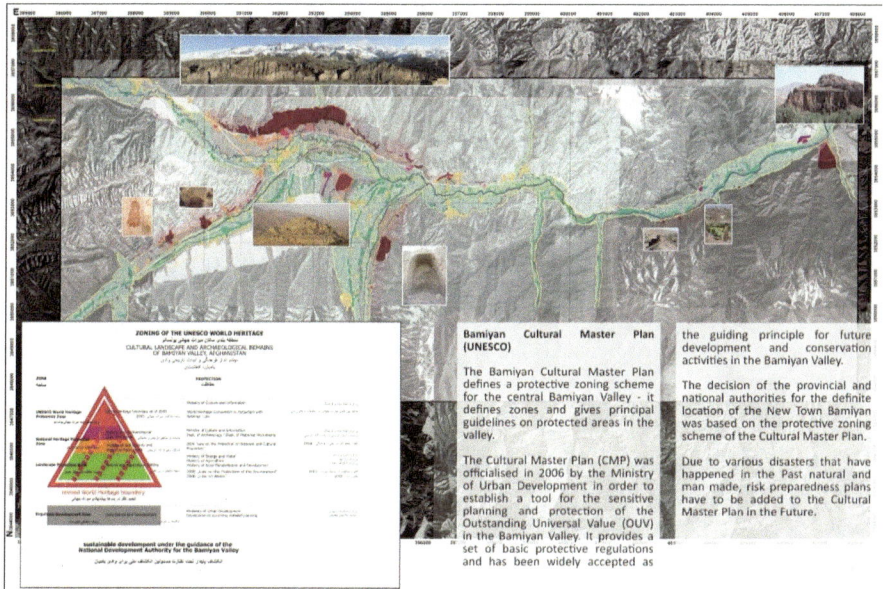

Fig. 4 The Bamiyan Cultural Master Plan—the protective zoning scheme (*Source* G. Toubekis/ RWTHacdc)

3.2 The Protective Zoning Scheme of the Cultural Master Plan (CMP)

An essential component of the master plan is the zoning concept, illustrated as a colored pyramid-shaped scheme (Fig. 4), designating specific zones (CMP Zones 1–4) with varying levels of protection requirements, and identifying areas for coordinated urban development (CMP Zone 5). Adapting the boundary contour lines from the original 2003 large scale nomination map was necessary to make boundary lines identifiable on the ground along real-world features like roads and landscape segments.

CMP Zone 1—*UNESCO World Heritage Protection Zone*, with boundaries as defined in the initial application dossier for the nomination (red color).
CMP Zone 2—*Archaeological Heritage Protection Zone* featuring significant archaeological assets both above and below ground surface[11] (pink color).
CMP Zone 3—*National Heritage Protection Zone*, including traditional settlements, religious shrines, and vernacular architecture, constituting significant elements of the cultural landscape (orange color).

---

[11] As identified based on thorough archaeological survey and excavation by NRICP Japan.

Fig. 5 Documentation of traditional building techniques and vernacular architecture elements in the Cultural Master Plan (*Source* Pierre Smars, Daniel Lohmann / RWTHacdc)

CMP Zone 4—*Landscape Protection Zone*, covering the river's high-flood plain, irrigation systems, traditional agriculture, and the natural environment, such as nearby mountainous pastures (green color).

CMP Zone 5—*Regulated Development Zone*, addressing urban disturbances and development pressures while allowing for controlled urban development, subject to detailed planning (grey color).

Propagation of traditional building techniques. Managing a protected cultural landscape, unlike archaeological sites primarily characterized by visible monumental remains, necessitates active community involvement. This participation is crucial to prevent the loss of historic fabric due to modernization and to adapt land use practices to protect subsurface monuments not visible from above. Traditional building footprints regarding size, scale, and materials like adobe/wood offer a sustainable approach for settlement expansion. Restoring iconic structures, like shrines, preserves both traditional customs and historic relics. The CMP strongly promotes sustainable land use practices, preserving traditional irrigation systems and water management techniques (Fig. 5).

Topographic Map. The maps provided with the nomination were outdated and insufficient for effective planning purposes. Given the challenging terrain and resource constraints for field missions, the use of remote sensing technologies has proven to be instrumental in obtaining accurate information about the built environment and topography, and in acquiring precise three-dimensional spatial information about the Bamiyan area. A Digital Elevation Model (DEM) was generated using high-resolution stereo satellite imagery and photogrammetric analysis, covering a 22 × 7.2 km area with an elevation range from 2353 to 3130 m. Differential GPS (DGPS) measurements improved accuracy to 0.5 m. This DEM, combined with data from the CMP, provided a precise depiction of the ground situation, including the documentation of extensive archaeological areas within the Cultural Landscape reaching beyond the initially marked World Heritage property boundaries. A Geographical Information System (GIS) was created to store all data and plan materials, assist in plan implementation, and inform future decisions (Fig. 6).

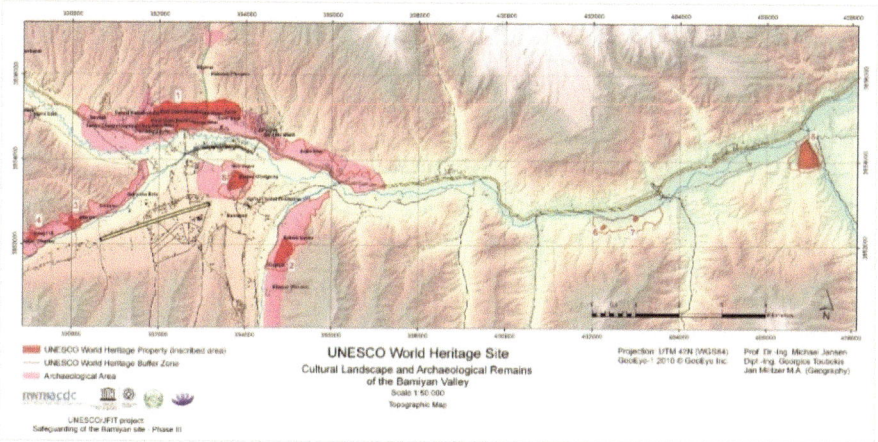

Fig. 6 Topographical map of the Bamiyan Valley; areas in pink, larger than the initial buffer zones, indicate a high probability of underground archaeological remains (*Source* G. Toubekis/RWTHacdc)

Monitoring Development. A pressing concern arises from the necessity to reconsider UNESCO buffer zones and their integration with local communities. This urgency is underscored by the unintended consequences of buffer zone delineations, originally established in sparsely inhabited areas, now intersecting with populated villages, and inadvertently delineating entire neighborhoods. This has given rise to misperceptions and discontent among the residents, necessitating a strategic re-evaluation of management practices. Satellite imagery is invaluable for monitoring the dynamics of urban growth, especially in the context of the rapid urbanization that ignored the proposals from the Cultural Master Plan and now threatens the property's Outstanding Universal Value (Fig. 7).

Virtual Technologies for Reconstruction. The cultural master plan includes digital simulations of the destroyed Buddha figures using virtual reality technologies. This innovation democratizes the general debate on the potential reconstruction of the site and the Buddha figures by involving both experts and the public. These virtual models will serve as communication and planning tools, enabling collaborative envisioning of future consolidation measures and discussions about the future of the World Heritage Site as a whole. These digital representations integrate detailed structural findings from high-resolution surveys and spatial configurations around the vacant rock niches. Using virtual reality, the plan promotes informed decision-making and public participation in shaping the site's destiny (Toubekis et al. 2021, 2020; Toubekis and Jansen 2013).

Fig. 7 Change detection—a comparison from 2004 to 2010 from different satellite images (*Source* Quickbird/GeoEye Inc.)

4 Lessons Learned—Balancing Tourism Development, Community Engagement, and Heritage Reconstruction

The Bamiyan case study highlights resilience during conflict in pursuit of sustainable development goals. The management of World Heritage sites requires cooperation between international, national, and local stakeholders. Adaptive risk assessment and disaster preparedness are essential, considering evolving environmental and social factors. World Heritage principles can enhance the environmental and social sustainability of cultural landscapes by integrating traditional land management and community participation. Future planning for the Bamiyan Valley should integrate envisioned cultural development projects, such as the archaeological park, into regional development, encompassing the new city and cultural landscape with its heritage elements. Bamiyan demonstrates the multiple values of cultural heritage in multicultural societies, going beyond preservation to support broader development goals, including identity, peace, reconciliation, and international cooperation, which resonate globally, especially with younger generations.

Empowering Communities in Cultural Heritage Preservation. The deliberate destruction of the world's largest Buddhist statues highlights the importance of careful planning and management of UNESCO World Heritage Sites. Preserving these sites extends beyond monument protection, encompassing the definition of associated values. Transparent negotiations and integrated restoration efforts can aid peace processes, especially in contexts of intergroup mistrust. Cultural heritage preservation becomes a tool for peace, highlighting the link between attacks on

Fig. 8 Bamiyan Stakeholder meeting moderated by UNESCO Kabul office in April 2017 and Prof. Michael Jansen and his team from the Research Center Indian Ocean (RIO) (*Photo* Hiromi Yasui)

heritage, human rights, and overall security. Strengthening international legal instruments, like the Hague Convention and UNESCO Conventions, is crucial in combatting illicit cultural property trafficking. The Bamiyan cultural master plan emphasizes the importance of respecting the identity of affected communities during heritage site preservation, especially in post-crisis scenarios. Involving local communities and listening to their views and also their doubts is vital for achieving planning goals, especially ensuring the authenticity of reconstruction efforts. This can only be achieved through open and inclusive discussions on the planned interventions, allowing the generation still affected by the losses of conflict to actively engage and participate (Fig. 8).

Exploring Tourism Opportunities. Preserving a cultural landscape requires regulating both road and air traffic. Therefore, the Bamiyan Cultural Master Plan suggests relocating the valley's airfield and redirecting major roads away from archaeological sites would create new development areas at a safe distance from cultural elements. Sustainable use of cultural heritage entails promoting national and international tourism for cultural exchange. Bamiyan has investigated multiple forms of tourism, including the Silk Road Music Festival, eco-friendly tourism at Band-e-Amir Lakes, and community-based winter tourism in the alpine highlands with international aid (Ashley and Dear 2011). Prioritizing generating revenue from domestic tourism can potentially boost the country's economy. The residents of Bamiyan have effectively utilized tourism to promote the diverse identities of their multiple communities and encourage the respectful preservation of Bamiyan's cultural heritage, benefiting both local and global visitors. The goal is to reap economic, social, and cultural benefits from tourism while avoiding the detrimental commercialization of cultural values and assets (ICOMOS 2022).

Capacity Building for Sustainable Development and Progress. Capacity building is essential for the sustainable conservation of cultural landscapes. To make meaningful progress, we must invest in initiatives that equip local communities and stakeholders with the knowledge and skills necessary for successful landscape management. A full understanding of UNESCO's cultural landscape concept is crucial to the effective implementation of the mandate of the World Heritage Convention. The Bamiyan Cultural Master Plan has improved this understanding by involving government agencies and local stakeholders, and by making concepts accessible in different local languages. Public consultation on heritage issues improves governance, particularly where citizen-authority relations are not effective and inclusive. Prioritizing training and human resources bridges knowledge gaps across generations and cultures, facilitating knowledge sharing and inclusive decision-making. These investments support sustainable heritage conservation practices, including the preservation of underground archaeological remains.

5 Conclusion and Outlook

In the 21st century, cultural heritage management now extends beyond monuments to include the environment and people's relationships with their surroundings, emphasizing capacity building, empowerment of local communities, and fostering ownership and resilience, while promoting sustainable tourism and sustainable development. Management of cultural heritage requires the involvement of all stakeholders, particularly in post-conflict situations where their participation is essential for reconstruction and achieving sustainability objectives. The Florence Conference provides inspiration and an opportunity to recall specific European contributions to the heritage debate that can guide efforts in the current context of Afghanistan.

Recognition in Peacebuilding and Democracy. Recognizing the role of cultural heritage in promoting tolerance, reconciliation, and intercultural dialogue is essential. However, it can also be manipulated during conflicts, becoming a target or a catalyst for fighting and misinforming. Recognizing this duality is crucial in developing strategies to protect cultural heritage in volatile contexts. Europe has recognized the significance of cultural heritage in conflicts and crises, making it a component of peace and security in the European Union's external actions approach (Council of the European Union 2021; European External Action Service 2021). Drawing from Europe's profound experience and expertise in shaping cultural heritage and cultural landscapes for the 21st century is essential. The *European Landscape Convention*,[12]

---

[12] The protocol in 2016 (CETS no. 219) amended the convention allowing its formal adoption also by non-European States and changing the original title (Council of Europe Landscape Convention).

adopted in Florence (Italy) in 2000 under the Council of Europe, emphasizes landscape protection, management, and planning for societal well-being, involving more than physical features, and including transformative processes and conceptualization of landscape. The *Faro Convention (Convention on the Value of Cultural Heritage for Society)*, established in 2005, highlights cultural heritage's role in human rights, democracy, and social cohesion. Both conventions promote inclusive and people-centered approaches to heritage preservation, aligning with sustainable development objectives (Council of Europe 2017, 2005).

ICOMOS expertise and practice. Ensuring professional heritage preservation expertise within EU policy and aid programs the EU Commission and ICOMOS have developed a *Handbook on European Quality Principles* to guide stakeholders, emphasizing shared responsibility for cultural heritage and outlining key concepts, international standards, and the benefits of quality assurance mechanisms throughout the project cycle (Selfslagh and Rourke 2020). It acknowledges participatory planning involving civil society as a powerful tool for sustainable development and peace. Commitment, collaboration, capacity building, and meaningful connections are essential for success, transcending differences, and promoting a harmonious future. The *ICOMOS Florence Declaration Heritage and Landscape as Humans Values* highlights the importance of cultural heritage and landscapes in promoting democratic societies. It urges collective efforts to preserve and promote these values for positive social change through meaningful dialogue among diverse cultures (ICOMOS 2014).

Action is paramount. The preservation of cultural heritage is interconnected, and conflicts within diverse societies can have global repercussions, exacerbated by climate change. A comprehensive, sustainable development approach that integrates cultural heritage into spatial planning is essential and consistent with international law and policy. Success depends on cooperation and inclusiveness, which means promoting heritage conservation, conflict prevention, and sustainability simultaneously. However, this requires commitment, building capacity, and making meaningful connections across initiatives and across differences. At its core, it promotes sustainable, people-centered conservation and enhances the role of cultural heritage as a driver of societal change through extensive education, training, and long-term interdisciplinary partnerships.

Acknowledgements I express my gratitude to the late Professors Michael Jansen and Kosaku Maeda for their contributions to the scientific exploration and preservation of the Bamiyan valley and its cultural landscape. Their expertise, dedication, and commitment have influenced the field of heritage management in this region of the world, enhancing our understanding and appreciation of Bamiyan's cultural heritage. Their groundbreaking research will serve as a valuable guide and source of inspiration for future efforts in both research and conservation.

References

Ashley L, Dear C (2011) Ski Afghanistan: A backcountry guide to Bamyan & Band-e-Amir. Agha Khan Foundation, Afghanistan

Bumbaru D, Burke S, Petzet M, Truscott MC, Ziesemer J (eds) (2000) Heritage at risk. Patrimoine en péril. Patrimonio en peligro: ICOMOS world report 2000 on monuments and sites in danger. Heritage at Risk vol 1. Saur, München

Council of Europe (2005) Council of Europe framework convention on the value of cultural heritage for society-Faro Convention (CETS 179). CETS 199

Council of Europe (ed) (2017) Landscape dimensions. Reflections and proposals for the implementation of the European Landscape Convention. Council of Europe Publishing, Strasbourg

Council of the European Union (ed) (2021) Council conclusions (9837/21) on EU approach to cultural heritage in conflicts and crises

Emmerling E, Petzet M (eds) (2016) The giant Buddhas of Bamiyan II. Safeguarding the remains 2010–2015. Monuments and sites p 21. Bäßler, Berlin

European External Action Service (2021) 9962/21-Cultural heritage in conflicts and crises. A component for peace and security in European Union's external action

Francioni F, Gordley J (eds) (2013) Enforcing international cultural heritage law. Oxford University Press

ICOMOS (ed) (2014) ICOMOS florence declaration on heritage and landscape as human values

ICOMOS (ed) (2017) ICOMOS Guidance on post trauma recovery and reconstruction for World Heritage Cultural Properties

ICOMOS (ed) (2022) ICOMOS international charter for cultural heritage tourism: reinforcing cultural heritage protection and community resilience through responsible and sustainable tourism management

Jansen M (2009) Presentation of the cultural master plan Bamiyan (Kabul, 31 July/Bamiyan, 2 August 2006). In: Petzet M (ed) The giant Buddhas of Bamiyan. Monuments and sites, vol 19, pp 123–124. Bäßler, Berlin

Jansen M, Toubekis G (2020) The cultural master plan of Bamiyan: the sustainability Dilemma of protection and progress. In: Nagaoka M (ed) The future of the Bamiyan Buddha statues: heritage reconstruction in theory and practice. Springer International Publishing, Cham

Jansen M, Toubekis G (2007) Bamiyan cultural master plan-reports from field survey 2005–2007 within phase II of the UNESCO/JFIT project-safeguarding of the Bamiyan site. Aachen Center for Documentation and Conservation (ACDC), Aachen

Klimburg-Salter D (2020) Entangled narrative biographies of the colossal sculptures of Bāmiyān: heroes of the mythic history of the conversion to Islam. In: Nagaoka M (ed) The future of the Bamiyan Buddha statues: heritage reconstruction in theory and practice. Springer International Publishing. https://doi.org/10.1007/978-3-030-51316-0

Machat C, Ziesemer J (eds) (2020) Heritage at risk-ICOMOS world report 2016–2019 on monuments and sites in danger. Bäßler, Berlin

Manhart C (2001) The Afghan cultural heritage crisis: UNESCO's response to the destruction of statues in Afghanistan. Am J Archaeol 105:387–388. https://doi.org/10.2307/507361

Nagaoka M (ed) (2020) The future of the Bamiyan Buddha statues: heritage reconstruction in theory and practice. Springer International Publishing. https://doi.org/10.1007/978-3-030-51316-0

Petzet M (ed) (2009) The giant Buddhas of Bamiyan. Safeguarding the remains. Monuments and sites p 19. Bäßler, Berlin

Rapoport A (1992) On cultural landscapes. Traditional dwellings and settlements review vol 3, pp 33–47. International Association for the Study of Traditional Environments (IASTE)

Sabbioni C, Brimblecombe P, Cassar M (2010) The atlas of climate change impact on European Cultural Heritage: scientific analysis and management strategies. Anthem Press

Selfslagh B, Rourke G (2020) European quality principles for EU-funded interventions with potential impact upon cultural heritage. In: Selfslagh B, Rourke G (eds) Revised version. ICOMOS International, Paris

Silverman H, Ruggles DF (eds) (2007) Cultural heritage and human rights, pp 3–29. Springer, New York

Toubekis G, Jansen M (2013) The giant Buddha figures in Afghanistan: virtual reality for a physical reconstruction? In: Falser M, Juneja M (eds) "Archaeologizing" heritage? Transcultural entanglements between local social practices and global virtual realities, pp 143–166. Transcultural Research–Heidelberg Studies on Asia and Europe in a Global Context. Springer, Berlin

Toubekis G, Jansen M, Jarke M (2020) Physical revitalization of the Eastern Buddha statue in Bamiyan using reinforced adobe material. In: UNESCO (ed) The future of the Bamiyan Buddha statues: heritage reconstruction in theory and practice, pp 307–329. Springer International Publishing, Cham. https://doi.org/10.1007/978-3-030-51316-0

Toubekis G, Jansen M, Jarke M (2021) Cultural master plan Bamiyan (Afghanistan)–a process model for the management of cultural landscapes based on remote-sensing data. In: Ioannides M, Fink E, Cantoni L, Champion E (eds) Digital heritage. Progress in cultural heritage: documentation, preservation, and protection: 8th international conference, EuroMed 2020, virtual event, November 2–5, 2020, revised selected papers, pp 115–126. Information Systems and Applications, Incl. Internet/Web, and HCI. Springer International Publishing. https://doi.org/10.1007/978-3-030-73043-7

UNESCO (1972) Convention concerning the protection of the world cultural and natural heritage

UNESCO World Heritage Centre (ed) (2014) Climate change adaptation for natural world heritage sites–a practical guide. World heritage papers series 37. UNESCO

UNESCO World Heritage Centre (2012) Safeguarding of the Bamiyan site, phase IV. https://whc.unesco.org/en/activities/717/. Last Accessed 22 Oct 2023

United Nations Assistance Mission in Afghanistan (UNAMA) (ed) (2022) Human rights in Afghanistan: 15 August 2021 to 15 June 2022. UNAMA, Kabul

Cultural Landscape in the Bamiyan Strategic Masterplan 2018

Mirella Loda⊙ **and Gaetano Di Benedetto**

Abstract The paper deals with the approach taken for the protection of the cultural landscape in the Bamiyan Strategic Masterplan. It starts by illustrating the evolution of the concept of cultural landscape in the UNESCO documents and the difficulty faced in implementing the *Bamiyan Cultural Master Plan* (BCMP) of 2004–13. It goes on to describe the measures contained in the *Bamiyan Strategic Masterplan* (BSMP) of 2018 that are specifically aimed at safeguarding the landscape. Special focus is placed on avoiding population and settlement increases on the valley floor and the bypass road cutting across the valley floor.

Keywords Cultural landscape · Bamiyan world heritage site · Strategic Masterplan

1 Premise

The Bamiyan Strategic Masterplan 2018 (BSMP) (LaGeS 2018) was prepared in 2017–18 as part of a cooperation project between the University of Florence and the Afghan Ministry for Urban Development and Housing (MUDH) and funded by the Italian Agency for Development Cooperation. The BSMP was officially adopted on November 14, 2018.

The goal of a Strategic Masterplan is to outline the best medium to long term solutions for the sustainable use and enhancement of regional resources. In the case of Bamiyan, owing to its inclusion on the UNESCO list of world heritage in danger (since 2003), the safeguarding and enhancement of cultural heritage was one of the primary points of focus during the preparation of its Strategic Master Plan.

According to its nomination dossier (2003), Bamiyan's cultural heritage consists of exceptional "cultural landscape and archaeological remains". However, while

M. Loda (✉) · G. Di Benedetto
Department of History, Archaeology, Geography, Fine and Performing Arts (SAGAS) and
Laboratory for Social Geography (LaGeS), University of Florence, Via San Gallo 10, 50129
Firenze, Italy
e-mail: mirella.loda@unifi.it

M. Loda and P. Abenante (eds.), *Cultural Heritage and Development in Fragile
Contexts*, Research for Development, https://doi.org/10.1007/978-3-031-54816-1_6

putting Bamiyan on the UNESCO World Heritage List has yielded important results concerning the protection of the archaeological remains, the recovery of wall paintings and the safety of the niches which held the Buddhas,[1] there emerged much uncertainty regarding the cultural landscape and—beyond generic restrictions—it is not known which measures should be taken, and which bodies should be appointed, to safeguard the cultural landscape.

Fundamental questions remained open, such as how much of the area should be safeguarded; what should the transition from a description of the fundamental components of the cultural landscape to a safeguard plan look like; and how to find a balance between the protection of the cultural landscape and the local population's legitimate calls for modernization and development.

Starting from here, this paper will first examine and comment on the reasons for the delay in landscape protection interventions, and then illustrate how the topic of landscape protection was dealt with in the Bamiyan Strategic Master Plan (BSMP) 2018.

2   Cultural Landscape: A Complex Concept

As for the delay in interventions for the safeguarding of the cultural landscape, one reason for this can certainly be found in the complexity of the very concept of landscape, and in the notion of landscape targeted by UNESCO action (Loda and Pettenati 2023): very evocative, as we will see, but hard to transform into an effective protection plan.

This conception of landscape derives from that developed in French regional geography at the start of the twentieth century (and popularized in the United States by Carl Sauer) while analyzing the landscapes forged by traditional (European) farming civilizations.

The UNESCO approach to the topic of landscape is outlined in three fundamental documents, which also provide the basis for the following operational guidelines: the "Recommendation Concerning the Safeguarding of the Beauty and Character of Landscapes and Sites" from 1962 (UNESCO 1962); the "Convention Concerning the Protection of the World Cultural and Natural Heritage" from 1972 (UNESCO 1972) which laid down the World Heritage Convention and the relative Heritage List; and the "Budapest Declaration" from 2002 (UNESCO 2002a).

In the *Recommendation*, which laid down the first cornerstone for the protection of cultural heritage of universal value and the progressive construction of the World Heritage List (WHL), the term landscape never appears alone but always alongside the word "site". Hence, the purpose of the conceptual device "landscape and site" is to place the specific object of interest (the site) in a fundamentally aesthetic frame which—by referring to an idea of beauty and harmony, unity and balance between

---

[1] This is thanks to the commitment of many experts and the resources that have been made available by donor countries.

man and nature, and stability—exalts its meaning and intrinsic value. At the same time the interchangeable use of the terms "site" and "landscape" projects onto the latter the idea of a physical entity, to be delimited and preserved with the same principles applied to the preservation (or museumization) of a site.

Article 1 of the 1972 *Convention* identifies cultural heritage as "…groups of separate or connected buildings which, *because of* their architecture, their homogeneity or *their place in the landscape,* are of outstanding universal value from the point of view of history, art or science" (UNESCO 1972) (our italics). Again, the landscape, understood reductively as a portion of physical space where single human buildings are situated, is used to refer to an idea of a whole, a setting, a (harmonious) context that substantiates the exceptional value of a particular heritage. In Article 1, the *Convention* also provides a definition of "site" which would reappear later on as the definition of "cultural landscape": "Sites: works of man or the combined works of nature and man, and areas including archaeological sites which are of outstanding universal value from the historical, aesthetic, ethnological or anthropological point of view" (UNESCO 1972).

The "cultural landscape" concept appears for the first time in the 1992 version of the *Operational Guidelines for the Implementation of the World Heritage Convention,* (UNESCO 2021) that is, twenty years after the *Convention* itself,[2] to then be reused in all the following updates. The concept adopts the idea of a whole, a specific context, which already belongs to the concept of landscape, while strengthening the sense of continuity with a historical-cultural construct and a universe substantially inside traditional farming societies. Indeed, as designated in Article 1 of the *Convention,* cultural landscapes are the result of "combined works of nature and of man". "They are illustrative of the evolution of human society and settlement over time, under the influence of the physical constraints and/or opportunities presented by their natural environment and of successive social, economic and cultural forces, both external and internal" (Brown 2018, § 47). In the same way, under "Cultural Landscapes", the UNESCO homepage indicates "cultivated terraces on lofty mountains, gardens, sacred places …—testify to the creative genius, social development and the imaginative and spiritual vitality of humanity. They are part of our collective identity."

Lastly, the *Budapest Declaration* (2002) also proposes a similar formulation, hinging on an almost interchangeable use of the terms site and landscape, but where the second term more explicitly takes on the meaning—very close to the meaning assigned to it in French regional geography—of the visual projection of a particular cultural context: "We care for these sites from the deepest forests to the highest mountains, from ancient villages to magnificent buildings, so that the diverse landscapes and cultures of the world be forever protected".

None of the three texts contains a definition of landscape. The term landscape never appears on its own but is always in association with the term "site". By putting "landscape and site" together, the landscape is reduced to the aesthetic frame that

---

[2] Interesting reflections on the historical background of the meeting of the World Heritage Committee which took place at La Petite Pierre (France) in 1992 are found in Brown (2018).

contains the actual object of interest, i.e., the site, whose intrinsic value is underlined by conveying an idea of a balance between humankind and nature.

Hence the establishment and then the management of the landscape heritage of humanity took its cue from the aesthetic and therefore fundamentally visual sense that the term landscape takes on in everyday language.[3] As such, the WHL was able to absorb all its different meanings and evocative power, to go beyond an idea of heritage as a collection of single events/assets and read their value in contextual terms. At the same time, it inevitably paid for the elusive nature of this conception of landscape which makes its translation in operational terms—i.e., its identification, delimitation, protection and management—extremely complex and fragile.[4]

The criticalities resulting from the adoption of quite lax basic definitions in UNESCO practices relating to landscape heritage have moreover been accentuated by the need to satisfy the criteria of authenticity and integrity[5] which have been a condition for nomination on the WHL. What is more, it has augmented the already arduous task of identifying shared and so-to-speak transcultural criteria to distinguish between the categories of cultural and natural landscape.[6]

Among these issues, subject to broad debate on an international level, I would like to recall the following problems, owing to the significance they take on in the case of Bamiyan (Afghanistan) presented in this article:

– The difficulties connected with identifying the constitutive elements and defining the boundaries of a cultural landscape;
– The unstable balance between protection of the landscape and transformation of the society and territory.

In terms of the UNESCO goals, despite the explicit mention of the cultural landscape in the title of the nomination document, the attention and protective actions following Bamiyan's inclusion on the list concentrated exclusively on the archaeological remains, which were rapidly marked off along with the protection areas. The protection of the cultural landscape, on the other hand, found itself relegated to an afterthought.

Between 2005 and 2007, the Technical University of Aachen, supported by UNESCO, carried out a survey of the territorial resources which, as well as adding to the census of archaeological heritage, enabled the recognition of important elements

---

[3] On this topic, see the interesting research from the 1970s in which Gerhard Hardt demonstrated how the modern concept of "landscape" corresponds to the idea of "beautiful scenery" (Hardt 1970). In effect, the abstract idea of landscape is associated with concepts such as "harmony", "totality", "context" and "synthesis". Hence, it defines a semantic field that corresponds to an abstract ideal of beauty applied to the landscape, originating in the context of bourgeois cultural tradition.

[4] According to Fowler (2003, p. 42), despite having the requirements, many sites do not request registration on the cultural landscape list owing to the definition and management difficulties that this would imply. The same reflection was made by Fowler in his contribution to the UNESCO conference on cultural landscape held in Ferrara in 2002 (UNESCO 2002b).

[5] On the elusivity of the concept of integrity and the problems connected to its evaluation, especially in rural settings (Gullino and Larcher 2013).

[6] As shown by the recent appearance of hybrid concepts such as "CultureNatures" (ICOMOS 2017).

of the landscape and offered a detailed description of the geographical and physical features of the territory, the settlement and architectural structures, water resources and water regulation systems (ACDC 2013).

Nevertheless, the resulting *Bamiyan Cultural Master Plan* (BCMP)—beyond the ambitious title—was more a stock-take of landscape elements[7] than a real planning tool. The upshot was that it had no influence in terms of standards to protect and enhance the cultural landscape. Thus, the only regulatory instrument available for use in the Bamiyan context remained the national "Afghan Law on the Protection of Historic and Cultural Properties" of 2004 for the protection of monuments (not landscape), which was improperly taken as the benchmark for landscape protection in buffer zones.

Moreover, the elements identified by the BCMP as making up the local cultural landscape are distributed in a vast area of around 32 km^2, equivalent to the whole valley floor. To identify this entire area as UNESCO property could be quite problematic. At the same time, the UNESCO property cannot be reduced to a portion of the valley floor territory, except by way of totally arbitrary and questionable operations of landscape analysis. The topic is highly sensitive owing to the consequences that this definition would have on the future development of the areas falling either inside or outside the UNESCO property perimeter.

In sum, we can conclude that 20 years after Bamiyan's inclusion in the Heritage of humanity List, the action by UNESCO has without doubt been effective in protecting and promoting its archaeological heritage. On the landscape level, however, there has been a lack of clarity concerning plans and measures to be adopted.

3   The Approach of the Bamiyan Strategic Master Plan to the Protection of Cultural Landscape

The concept of cultural landscape has been subject to further reflection by UNESCO within the framework of the *Recommendation on the Historic Urban Landscape* (HUL) (UNESCO 2011).

This approach to cultural landscape has greatly helped overcome the impasse in defining the management of Bamiyan's cultural landscape while developing the Bamiyan Strategic Master Plan 2018. Although the concept of HUL has been developed mainly with regard to urban contexts, quite different from the Bamiyan valley, it is nevertheless an important step forward in discussions on landscape in general, because it clearly specifies, from a holistic and evolutionary perspective, that the central objective of any activity on landscape consists in integrating policies and

---

[7] According to Jukka Jokiletho the term "inventory" better describes the nature of the document (Jokilehto 2020, p. 201). Criticism of the "stock-take" approach to landscape values seeped into the town planning debate. In the last years of the twentieth century, this resulted in defining "structural invariants" as the baseline for outlining the landscape transformation rules. For a critical reflection on the concept of structural invariant and the difficulties accompanying its practical implementation in the Italian context (Maggio 2014).

practices of conservation of the built environment into the wider goals of (urban) development: "The historic urban landscape approach is aimed at preserving the quality of the human environment, enhancing the productive and sustainable use of urban spaces, while recognizing their dynamic character, and promoting social and functional diversity. It integrates the goals of urban heritage conservation and those of social and economic development. It is rooted in a balanced and sustainable relationship between the urban and natural environment, between the needs of present and future generations and the legacy from the past" (UNESCO 2011, § 11). Attention is paid to 'preserving the quality of the human environment (…) while recognizing [its] dynamic character and promoting social and functional diversity'.[8]

This approach to cultural landscape incorporates the results of the lively scientific debate on the issue, according to which the landscape is both a 'product and process' at the same time; it is a complex set of continually evolving tangible and intangible elements. From this perspective, in the context of shifting landscapes, the goals of heritage conservation are to be integrated with those of social and economic development. This will allow for the encompassing of gradual changes in the cultural landscape, resulting in a continuing and lasting transformation, physically, socially and symbolically.

So the landscape is the visible outcome of socio-economic and cultural processes, of underlying (production) structures that govern the transformation of the territory.[9] The attention and protective actions should, therefore, concentrate on these and not on single visible elements that we see in the territory. In the case of Bamiyan, the above means reflecting on the farming practices that have forged the rural landscape on the valley floor. It is essential to ensure their feasibility and profitability so that the plots of land, buildings and cleaning of the channels—in short, taking care of the land—are not abandoned (Fig. 1).

We cannot overlook the fact that feasibility and profitability are connected with the dynamicity and evolution of the farming landscape. Bamiyan's rural landscape has undergone profound transformations over the course of recent history. Until the 1970s, the typical crop was wheat, while today the dominant crop is the potato. This change to a typical "cash crop" is rooted in the progressive integration of Bamiyan's agriculture into the national and international markets, recently facilitated by improvements in the road network, connecting Bamiyan with Kabul and Herat. The shift from wheat to potato cultivation, which calls for much more water, has played a decisive role in broadening the irrigation system that shapes the current rural landscape.

At the same time, the valley floor is also the portion of the territory that has been most affected by the transformative pressure of the typical urban functions in recent years—just think of the rate of expansion of the bazaar (Figs. 2 and 3). Only

---

[8] While on one hand UNESCO appreciates the "continuing landscapes", "closely associated with traditional ways of life" (Roessler and Lin 2018, p. 4), particularly if connected to the permanence of traditional crops, the transformative dynamics of farming practices pose very great challenges to application of the integrity principle (Gullino and Larcher 2013).

[9] We wish to stress the consonance between this approach and the reflections on landscape developed by Lucio Gambi in the 1960s (Gambi 1966).

Fig. 1 Bamiyan agricultural landscape (*Photo* M. Loda, 2017)

comprising around 50 shops in 2002, it has expanded with great speed since 2011. According to the comprehensive survey carried out by LaGeS in spring 2018 (LaGeS 2018), the bazaar currently consists of 1,487 permanent shops, as well as 566 mobile businesses, providing work for a lot of the inhabitants of the new settlements in the north of the valley who do not own land and seek employment in non-farming activities. Hence, it is mainly towards this area that measures to limit urbanization processes and safeguard the landscape must be directed.

Concerns also arise regarding the widespread changes observed in the farm buildings, the *qal'a*. These make up a typical element of Bamiyan's cultural landscape, and are often subject to new interventions which contrast starkly with traditional building techniques (Fig. 4).

These issues provoked widespread reflection during the preparation of the Bamiyan Strategic Master Plan in view of both safeguarding the cultural heritage and meeting the needs of the farmers, while promoting the region's economy.

With no explicit borders or definition of the cultural landscape to protect in the UNESCO documents, the BSMP put forward solutions based on the principles of

Fig. 2 Start year of shops (*Source* LaGeS (2018, p. 100))

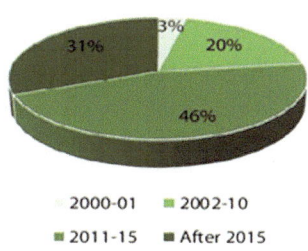

Fig. 3 Expansion of the new bazaar (*Source* authors' processing from satellite photos)

Fig. 4 Concrete interventions in a *qal'a* in Tolwara (*Photo* M. Loda 2023)

protecting the territory and on urban planning-type regulations. A first objective regarding land protection was to reduce consumption of the agricultural land on the valley floor as much as possible, as it constitutes a particularly valuable resource in a mountainous area such as that of Bamiyan, not to mention, a defining feature of the cultural landscape. Therefore, the BSMP identified the *cultural landscape* as the *agricultural landscape* of the whole valley floor.[10] The valley floor was categorized as a sort of extension of the UNESCO buffer zones that host the properties, but under the management of urban planning tools.

The first objective of the plan consisted of preventing the progressive urbanization of the valley floor and in safeguarding flat land farming. To this end, the BSMP developed a new tool, the "planning district", so as to issue different rules to the various parts of the municipality, according to their specific characteristics. The territory covered by the BSMP was divided into six planning districts, each governed by distinct development regulations.[11]

The whole valley floor was included in planning district no. 1, regulated by particularly rigid restrictions. In District 1, the plan does not allow for the settlement of additional inhabitants, nor the construction of additional dwellings to those already existing[12]; moreover, the plan does not permit the existing buildings to be raised, thereby protecting the views over the sites of archaeological interest. At the same time, by enabling the existing farms to expand their agricultural outbuildings, the BSMP aimed to safeguard the feasibility and profitability of the farming activities and avoid their abandonment, thus ensuring the continuation of land care practices (Fig. 5).

Farming on the valley floor and the agricultural landscape were also protected through the same solution produced with the aim of resolving the delicate question of vehicles crossing through the valley. A highly tense public debate had grown around this issue, between those wanting to protect the landscape and those calling for modernization. The bypass project drawn by the Ministry of Public Work and

---

[10] In a recent publication Jansen and Toubekis also agree on the need to clearly define the area that is to be protected—albeit at a late stage and curiously without ever quoting the BSMP: "The title of the World Heritage nomination 'Cultural Landscape and Archaeological Remains of the Bamiyan Valley' can be misleading, since the cultural landscape under the protection of the Convention is covering only a small fraction of the total cultural landscape" (Jansen and Toubekis 2020, p. 82). However, we should distinguish between the term agricultural landscape, which we have opted for, and rural landscape—put forward by the two authors in the text. While the former refers to the visible outcome of a practical economic activity (agriculture), the latter evokes the idea of non-urbanity that is difficult to reconcile with Bamiyan's recent development, and its taking on functions that are typical of a central area within the district and province.

[11] The BSMP divides the municipality of Bamiyan into six planning districts, each of which is designated an acceptable number of additional inhabitants, infrastructure and town planning regulations.

[12] The BSMP aims to limit the demographic expansion and consequent housing demand in the valley to around half of the increase forecasted up to 2037, while directing the remaining part towards the new city of Pasnaw, located on the strategic road that will connect Bamiyan with Shaidan and Shebartu (the site allocated for the new airport) (LaGeS 2018). Planning districts 2, 3, 4 and 5 have been earmarked to host the demographic increase in the valley.

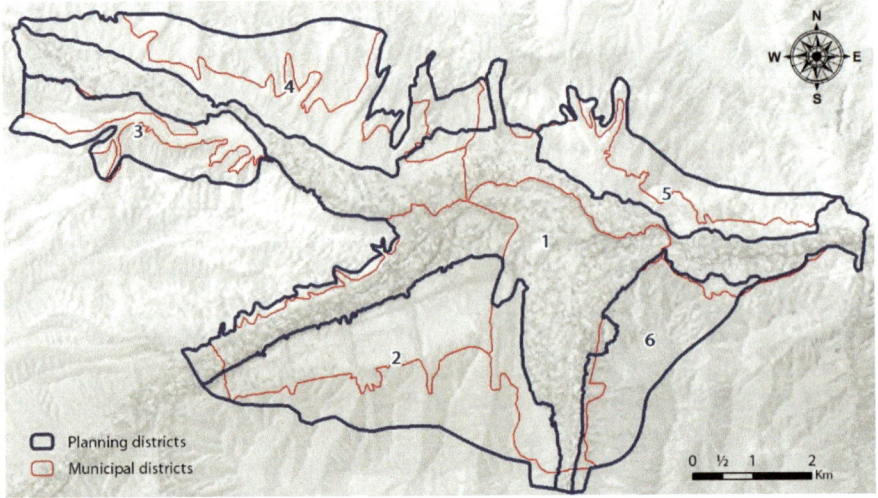

Fig. 5 Planning districts in the BSMP 2018 (*Source* LaGeS (2018, p. 164))

accepted by the Ministry of Urban Development and Housing as part of the (thank-fully never adopted) Bamiyan Urban Development Plan 2012, seriously challenged the safeguarding of the cultural landscape. By cutting across the valley floor a lot of farmland would have been swallowed up, and the portion of the valley around the UNESCO site of Gholghola would have been rapidly urbanized.

It is in fact quite surprising that the proposal had been accepted as consistent with the Bamiyan Cultural Master Plan (BCMP) by the UNESCO experts evaluating the Strategic Plan Bamiyan elaborated in 2013 by the Afghan Ministry for Urban Development (Fig. 6).

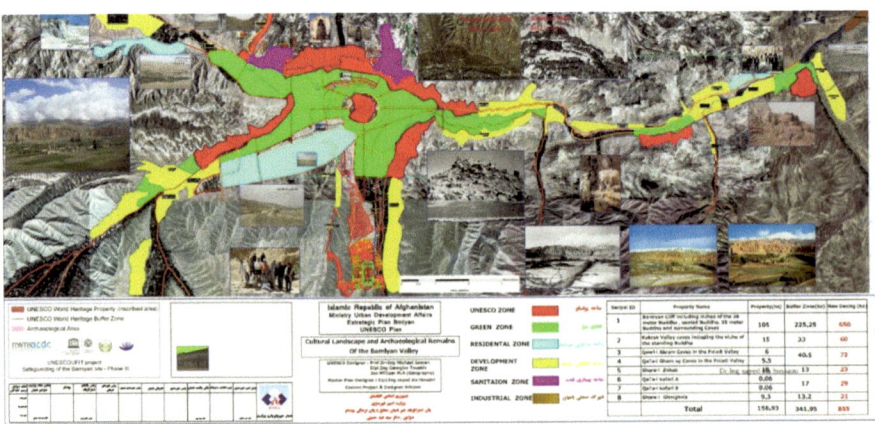

Fig. 6 Bamiyan urban development plan 2012 of the Ministry of Urban Development and Housing

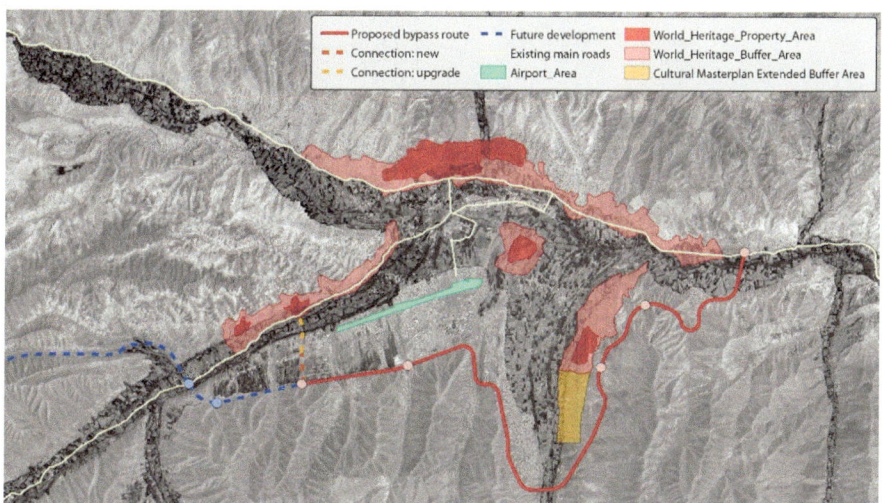

Fig. 7 Bypass proposed by the BSMP 2018 (*Source* LaGeS (2018, p. 190))

The drawing up of BSMP 2018 dealt with the matter of a variant to the ministerial Bypass project. The proposed solution frees up the valley floor by taking the road along the sides of the southern mountains until it joins the road for Kabul further east (Fig. 7). This solution was officially adopted as Bamiyan City Masterplan by local and national authorities on November 14, 2018 (Fig. 8).

4 Conclusions

The strategy adopted by the BSMP to protect the cultural landscape of the valley does not delve into the question of which structural invariants of the cultural landscape to protect, as per its inclusion in the WHL. This task is central to any Management Plan.[13] However, it provides a planning perspective for safeguarding the cultural landscape within the framework of the valley's general development, a perspective that was missing until that time. Moreover, in line with the international debate, the Bamiyan situation calls for a reflection on the exact definition and operationalization of the concept of cultural landscape that should be adopted in future interventions. The focus should shift from a vision of the landscape as a collection of elements with "Outstanding Universal Value" to be preserved towards the in-depth comprehension of the functioning of the socio-cultural and production system that this landscape has produced. Finally, greater and more focused attention should be placed on the requirements of the production system responsible for and guaranteeing the (re)generation and transformation of the landscape.

---

[13] The process to develop a Management Plan for Bamiyan has been started in April 2023.

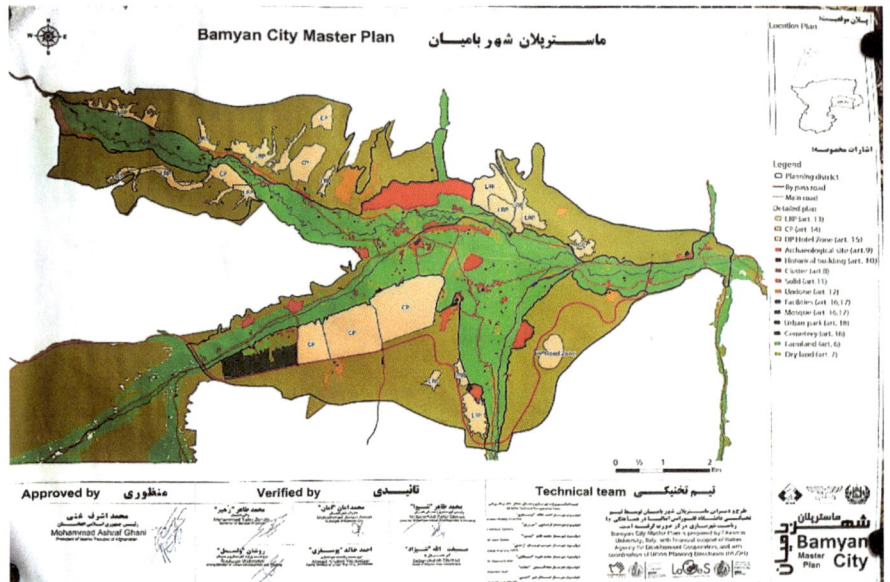

Fig. 8 Bamiyan city Masterplan

CRediT authorship contribution statement. Mirella Loda: Conceptualization, Methodology, Supervision, Formal Analysis, Writing—Original Draft, Visualization, Writing, Review & Editing. Gaetano Di Benedetto: Conceptualization, Methodology, Supervision.

References

ACDC (Aachen Center for Documentation and Conservation) (2013) Cultural masterplan Bamiyan, campaigns 2003–2007 (omnibus volume). Aachen University
Brown S (2018) World heritage and cultural landscape. Herit & Soc 11:19–43
Fowler P (2003) World heritage cultural landscapes 1992–2002. UNESCO, Paris
Gambi L (1966) Critica ai concetti geografici di paesaggio umano. Fratelli Lega, Faenza
Gullino P, Larcher F (2013) Integrity in UNESCO world heritage sites. A Comp Study Rural Landsc, J Cult Herit 14:389–395
Hardt G (1970) Die 'Landschaft' der Sprache und die 'Landschaft' der Geographen. Semantische und Forschungslogische Studien. Colloquium Geographicum 11, Ferd. Dümmler Verlag, Bonn
ICOMOS (2017) Learnings & committments from the culture nature journey@ the 19th general assembly. Dehl, https://www.icomos.org/en/77-articles-en-francais/49783-19th-gen eral-assembly-2017-results-of-the-scientific-symposium. Last Accessed 06 June 2021
Jansen M, Toubekis G (2020) The cultural master plan of Bamiyan: the sustainability Dilemma of protection and progress. In: Nagaoka M (ed) The future of the Bamiyan Buddha statues. Heritage reconstruction in theory and practice. Springer
Jokilehto J (2020) Reflections on the case of Bamiyan. In: Nagaoka M (ed) The future of the Bamiyan Buddha statues. Heritage reconstruction in theory and practice, pp 205–214. Springer-UNESCO, Paris, Cham

LaGeS (Laboratorio di Geografia Sociale, Università degli Studi Firenze) (2018) Bamiyan strategic master plan. Polistampa, Firenze

Loda M (2023) What cultural landscape for Bamiyan (Afghanistan)? Observations on the UNESCO site protection practices. In: Pettenati G (ed) Landscape as heritage. Routledge, London, pp 230–243

Maggio M (2014) Invarianti strutturali nel governo del territorio. Firenze University Press, Firenze

Roessler M, Lin RC-H (2018) Cultural landscape in world heritage conservation and cultural landscape conservation in Asia. Built Herit 3:3–26

UNESCO (1962) Recommendation concerning the safeguarding of the beauty and character of landscapes and sites, http://portal.unesco.org/en/ev.php-URL_ID=13067&URL_DO=DO_TOPIC&URL_SECTION=201.html. Last Accessed 03 June 2021

UNESCO (1972) Convention concerning the protection of the world cultural and natural heritage, https://whc.unesco.org/en/conventiontext/. Last Accessed 07 June 2021

UNESCO (2002) World Heritage Committee, Twenty-sixth session, The Budapest Declaration on World Heritage, https://whc.unesco.org/en/decisions/1217/. Accessed May 28, 2021.

UNESCO (2002) Cultural landscapes, challenges of conservation. Ferrara, Paris

UNESCO (2011) Recommendation on the historic urban landscape, https://whc.unesco.org/en/hul/#:~:text=About%20the%20Recommendation%20on%20the,in%20a%20changing%20global%20environment. Last Accessed 08 June 2022

UNESCO (2021) Operational guidelines for the implementation of the world heritage convention, https://whc.unesco.org/en/guidelines/. Last Accessed 08 June 2022

Cultural Heritage and Urban Development

Cultural Heritage and Urban Development: Embracing the Historic Urban Landscape Approach

Carlo Francini⊙ and Tatiana Rozochkina

Abstract In the global context, there is not a single historic city that has managed to preserve its original character intact; this concept is perpetually evolving, destined to transform in tandem with society itself. Given the constant urban change, how can the societies accommodate the need for urban development and the cultural heritage conservation within urban management? The article elaborates on the Historic Urban Landscape (HUL) approach as a solution framework that provides an understanding of how historic fabric and contemporary development can harmoniously interact, reinforcing their respective roles and significance. The case study of Florence serves as a compelling example of efficacy of this approach and offers valuable insights into the HUL's successful integration into existing local frameworks, ultimately shedding light on the potential benefits of the HUL application in different urban contexts.

Keywords Historic Urban Landscape · Urban policies · Integrated management

1 Historic Urban Landscape (HUL) and Sustainable Urban Development

The World Heritage List stands as a testament to the collective endeavor of humanity to preserve the treasures of the past for future generations. Among the diverse array of sites inscribed on the list, approximately one-third of the properties are comprised of historic cities and historic urban areas. This finding underscores the cultural significance of urban landscapes, their profound impact on the narrative of human history, and in shaping the collective memory of humanity. From the ancient metropolises

C. Francini (✉)
Municipality of Florence, Florence World Heritage and relations with UNESCO Office, Via Giuseppe Garibaldi, 7, 50123 Firenze, Italy
e-mail: carlo.francini@comune.fi.it

T. Rozochkina
University of Florence, HeRe_Lab–Heritage Research, joint laboratory University and Municipality of Florence, Via Della Mattonaia, 8, 50121 Firenze, Italy
e-mail: tatiana.rozochkina@unifi.it

of Mesopotamia to the medieval cityscapes of Europe and the vibrant bazaars of the Orient, these historic sites exemplify a profound sense of place, identity, and continuity that resonates with people from diverse cultural backgrounds.

As urbanization continues to sculpt the global landscape, the safeguarding of these living historical entities becomes an imperative task. Challenges that historic urban areas face today arise from the intricate interplay of factors such as rapid urbanization, population growth, the tensions between globalization and local development, evolving social dynamics, and pressing environmental concerns (Bandarin and Oers 2012). These contemporary challenges add layers of complexity atop the enduring historical tension between urban development and conservation, which calls for an integrated approach that transcends conventional conservation paradigms (Rodwell 2010). It is within this context that the concept of the Historic Urban Landscape (HUL) assumes paramount importance, heralding a new paradigm in urban development that harmoniously integrates heritage values with urban change.

The modern vision of urban conservation, characterized by its holistic approach to urban heritage, marks the culmination of a complex intellectual evolution originating from the specialized field of monuments conservation. While the formal discussion around the Historic Urban Landscape (HUL) approach was initiated by the UNESCO World Heritage Committee in 2003, with the Vienna Memorandum of 2005 (UNESCO World Heritage Centre 2005) further shaping its formulation, the early inklings of recognizing the holistic significance of urban heritage can be found in the conservation and management tradition developed in the late nineteenth century, which gained a renewed relevance in the aftermath of the modernist urban transformations.

In this pivotal era, pioneering thinkers such as Riegl (1982) and Sitte (1965) laid the conceptual groundwork for conceiving urban heritage as an interconnected whole. Riegl's theory of values sought to interpret the conservation of monuments through the lens of their inherent cultural and historical worth, propelling a shift away from a purely materialistic approach. Concurrently, Sitte's groundbreaking perspective on the city as a historical continuum highlighted the need to derive guiding models for the development of the modern city from its historical evolution. It is indeed in this phase that foundational concepts that later found resonance in modern charters and the HUL approach began to take shape.

As the twentieth century unfolded, the advent of modernism and the rapid pace of urbanization precipitated the erasure of historical layers, threatening cultural identities that had shaped cities for centuries. In response to the fervent drive for rapid modernization, the concepts of memory value, esthetic enjoyment, and collective responsibility for conservation gained renewed significance as essential counterbalances. This stark clash between the conservative vision of preserving historical urban fabric and the relentless pursuit of modernization laid bare a fundamental question: how do we reconcile the imperatives of conservation and development? Can a middle ground be found between the 'rigid' conservation of the historic city and its complete removal and replacement? The nuanced opposition engendered by this query continued to fuel theoretical developments throughout the twentieth century, with the ideas of renowned urbanists, such as Jacobs (1961), Mumford (1938), and

Lynch (1960) seeking a harmonious integration of conservation and development within the urban context. Within this intellectual crucible, the seeds of the HUL approach were sown, proposing a visionary paradigm that navigates the intricate terrain between heritage preservation and urban change.

The definition given in the recommendation on the HUL (UNESCO 2011) gives a thorough understanding of the holistic vision of heritage represented by the HUL idea and highlights some of its most important conceptual developments: broader understanding of urban space, moving beyond urban conservation as a specialized practice and its "operational" potential worldwide.

"The historic urban landscape is the urban area understood as the result of a historic layering of cultural and natural values and attributes, extending beyond the notion of 'historic centre' or 'ensemble' to include the broader urban context and its geographical setting. This wider context includes notably the site's topography, geomorphology, hydrology and natural features, its built environment, both historic and contemporary, its infrastructures above and below ground, its open spaces and gardens, its land use patterns and spatial organization, perceptions and visual relationships, as well as all other elements of the urban structure. It also includes social and cultural practices and values, economic processes and the intangible dimensions of heritage as related to diversity and identity."

This definition lays the foundation for a comprehensive and integrated approach to the identification, assessment, preservation, and management of the historic urban landscape within the broader framework of sustainable development. The fact that an international legal instrument operates with such expansive and profound concepts is remarkable and aptly reflects a paradigm shift from traditional conservation frameworks that prioritize individual structures, towards an all-encompassing appreciation of the urban landscape as a whole. If compared to the vision set in the 1964 Venice charter (ICOMOS 1964), a founding document on urban conservation that considers historic city as a setting for historic monuments rather than a heritage system, it can clearly be seen how long way the international understanding of urban space, historical city, and its present value have gone over the past half a century. More than that, the groundbreaking conceptual framework of HUL positions it as an attitude as to how cities develop, a "mindset", as it was defined by the Expert Meeting on HULs held at the UNESCO Headquarters in November 2008.

Understanding HUL as an attitude rather than a heritage category is especially important as it allows the international community to arrange for the semantic and conceptual differences that inevitably underlie any process as complicated as articulating a set of universal guiding principles for historic city management. Understood in these terms, the HUL approach sets itself as a global shared attitude towards the key concepts mobilized in the historic city conservation and management practices, which represents another groundbreaking development overlooked by the first international documents (Cameron 2008).

Another important aspect of HUL is that the concept marks a significant drift away from the confines of urban conservation as a specialized practice. Rather than considering conservation in isolation, HUL reimagines the preservation process within the expansive framework of urban management and development. This link between the two—a crucial step to the integrated approach to sustainable management of a "living" historic city—is traceable in two milestone documents of the second part of the twentieth century, namely the Declaration of Amsterdam (ICOMOS 1975) and the European Charter of the Architectural Heritage (Council of Europe 1975). Both of them place the conservation as an objective of city planning and put an emphasis on the necessity of considering areas of historic interest rather than historic buildings alone. Elaborating the idea, the 1976 Recommendation concerning the Safeguarding and Contemporary Role of Historic Areas (UNESCO 1976) also associates development to conservation, expanding its scope from urban areas to the outside of the city. Furthermore, the need to "preserve the traditional social fabric and functions of the historic areas" cited in the document as one of the goals of conservation is a clear indication of the management concern in the international conservation discourse.

These developments, later crystallized in the HUL concept, make the conservation informed by a concern for the future as much as for the past. Within the context of HUL, in fact, the term "historic" transcends its conventional association solely with age, transforming into a concept imbued with particular meanings and intrinsic values (Jokilehto 2010). This evolution not only underscores the dynamism of urban landscapes but also emphasizes their relevance to contemporary societies. Establishing a past–present continuity that beams the heritage up from its historical setting to the city as it appears today, HUL advances beyond the confines of the specialized conservation field, embracing a broader narrative that integrates heritage into the continuum of urban life, encouraging sustainable development while nurturing the authenticity and distinctiveness of cities.

2 The Florentine Experience: Adopting the HUL Methodology as Part of Urban Development Toolkit

The value-based understanding of the historic city provided by the HUL is today widely seen as the basis for a methodology used to govern the territory in an integrated and holistic manner. The 2011 Recommendation defines the concept of HUL as a structured methodology encompassing six "phases" and applicable through four "tools" to holistically and integrally manage the territory. By linking the Outstanding Universal Value—the key concept underlying the recommendation—to the relationship between the historic fabric and its surroundings, as well as to the functions acquired by the city over time, HUL provides guidance and direction for urban conservation, development and management of historic cities and their surrounding areas.

The experience of the Historic Center of Florence that will be discussed further serves as a case study of how research, additional levels of protection, and international recommendations can be incorporated and implemented within existing municipal regulations. This institutional linking of historic centers with the surrounding landscape is a practice that can be applied to various urban contexts, including sites not listed on the World Heritage List.

Due to its strategic location between the Arno River and the hill system, Florence has preserved over the centuries its naturalistic character, wherein the traditional agricultural landscape remains discernible, adorned with rural houses, historic villas, and religious structures. The context in which the historic center is located highly contributes to the value of the World Heritage site, with its integration with the surrounding landscape being cited as a condition of integrity in the statement of Outstanding Universal Value of the Historic Centre of Florence. Thus, through the implementation of the HUL methodology, the Municipality of Florence intended to work on a balanced conservation of the fragile landscape surrounding the city center to guarantee the integrity of the World Heritage site.

The 2011 Recommendation on the Historic Urban Landscape of Unesco reads: "present and future challenges require the definition and implementation of a new generation of public policies identifying and protecting the historic layering and balance of cultural and natural values in urban environments." With this in view, the Florence World Heritage and Relations with UNESCO Office of the Municipality of Florence has implemented a series of actions to adopt HUL as a guiding operational category in its site management model, including strategic projects, the definition of the buffer zone and enhancement of the governance model.

The journey towards the HUL methodology implementation began with the project "Firenze dal centro alle colline: Belvedere e percorsi panoramici" conceived in 2011–2012 (Bini et al. 2015a). This study, carried out by the Florence World Heritage and Relations with UNESCO Office in collaboration with the Department of Architecture of the University of Florence, was aimed at guiding urban transformations of the Historic Center of Florence in a way to ensure the integrity and enhancement of the city's image. The study resulted in the identification of 50 panoramic points with corresponding visual axes that contribute to the identity of the Historic Center, particularly in terms of preserving the integrity of the city's skyline.

The subsequent research for defining the Buffer Zone of the site (Bini et al. 2015b) was based on elaborating the data from the aforementioned study and the overlay of different protection levels (areas with landscape restrictions, protected buildings, archeological interest areas, parks, areas of particular green value, minor historic centers, rivers, etc.). The identified buffer zone is a 10,480 hectares area surrounding the World Heritage site, developed to provide an additional layer of protection of the immediate backdrop, principal vistas, and other structural and functional characteristics of the site.

In the case of Florence, the acknowledgment of the World Heritage site's buffer zone both by UNESCO[1] and the local planning mechanisms[2] marked the incorporation of the HUL concept into the main instruments of local territorial planning and regulation. This significant step not only contributed to safeguarding the integrity of the landscape and the skyline of the historic center but also contributed to the enhancement of the "ecological corridors" which are useful in countering urban sprawl in peripheral areas (Francini 2022; Francini 2017; Francini and Montacchini 2021).

Consistent with the aim of tangibly enhancing the relationship between the historic center and the hill system, the proposal to extend the perimeter of the World Heritage site was developed. This proposal, approved during the 44th extended session of the World Heritage Committee in Fuzhou, China, in July 2021, aimed to rectify a formal representation error in the site's map that had excluded the San Miniato al Monte complex, explicitly mentioned in the 1982 Statement of Outstanding Universal Value. The site boundary revision presented an opportunity to include not only the missing attribute but also an area characterized by gardens, avenues, and ramp systems— a veritable urban park designed for enjoying the city's greenery and panorama, connecting the historic center to the hill ecosystem.

Another "tool" that facilitated the implementation of HUL principles in the context of the Historic Centre of Florence is related to the revision of the governance and efficiency check systems. As through the "landscape" lens HUL immediately suggests the necessity for cross-sectoral action, the Municipality involved the Steering Committee in the system of governance of the site. The committee consists of stakeholders on the local, municipal, and regional levels,[3] thus incorporating the concept of landscape management in operational terms and enhancing the efficiency of the HUL implementation mechanism. Meeting twice a year, the committee is a valuable opportunity for the stakeholders to align the common goals and the vision that transcends the competencies of a single authority, similar to the way in which the idea of the Outstanding Universal Value of the site transcends the boundaries of the formally understood historic center.

The steering committee's primary responsibility is to oversee the update and implementation of the Management Plan. This classic instrument of urban governance is another opportunity for the effective implementation of the principles of the HUL. During the updating of this 2022 Management Plan (Municipality of Florence 2022), the HUL approach was used as one of the programmatic documents for determining the macro areas and projects selected within the action plan.

---

[1] The buffer zone was approved by the World Heritage Committee on July 6, 2015, during its 39th session in Bonn.

[2] The second midterm variant of the Structural Plan and Urban Planning Regulation, approved by the City Council on May 13, 2020, introduced new forms of protection in the areas designated as the "Core Zone" and "Buffer Zone"of the Historic Center of Florence.

[3] The Steering Committee for the Historic Centre of Florence comprises: Municipality of Florence; Tuscany Region; Ministry of Culture; UNESCO Office, Secretariat I; Superintendence of Archaeology, Fine Arts, and Landscape for the provinces of Florence, Pistoia, and Prato; Regional Directorate of Museums in Tuscany.

For what concerns the enhancement of the monitoring system, significant changes within the World Heritage site and its Buffer Zones are monitored through Heritage Impact Assessment (HIA), another implementation tool related to the HUL approach (Chiesi et al. 2020). This methodology builds on existing Environmental Impact Assessments (EIA) and Strategic Environmental Assessments (SEA) employed to mitigate the potential negative impacts of development on the OUV of a World Heritage site. HIA is a non-binding instrument that offers guidelines for assessing the impacts of changes on heritage values. In the Florentine context, a tailored HIA model developed for the historic center is incorporated into the 2022 municipal Operational Plan. This model serves both at political and technical levels to guide interventions, particularly those involving infrastructure projects and the repurposing of disused areas within the World Heritage site and its buffer zone.

In the past years, the World Heritage Center organized a series of events under the auspices of the World Heritage Cities Programme to celebrate the 10th Anniversary of the 2011 UNESCO Recommendation on the Historic Urban Landscape.[4] Bringing together the site managers of the UNESCO World Heritage cities, these events served as a catalyst for the wider adoption of the HUL approach.

The Florence World Heritage and Relations with UNESCO Office of the Municipality of Florence participated in several initiatives organized by the World Heritage Cities Programme of the World Heritage Centre (WHC 2020), including those related to the celebrations of the 10th anniversary of the HUL, thus contributing to the dissemination of the strategies implemented in Florence and the exchange of good practices. A Call for Action (WHC 2011) was issued on this occasion—to which the City of Florence also adhered—to raise awareness of the authorities and communities and to accelerate the incorporation of the approach at the bureaucratic level.

3   Conclusion

The discussed model for site management—by retaining values in a changing environment—is a unique example of HUL implementation in Italy that has an immense potential for rethinking the World Heritage sites management principles worldwide.

For historic cities, the UNESCO Recommendation on the HUL stands as an invaluable asset due to its dual role as both a comprehensive theoretical framework and a practical operational tool for urban management and conservation. However, the international community is still in the early stages of the exploration into the depths of the historic urban landscape's significance. According to the latest report on the implementation of the 2011 Recommendation (WHC 2023), 36% of member states

---

[4] The extensive list of events held from 2021 is a testimony to the significant impulse given by the World Heritage Center to promoting the HUL. Of particular note is the summary report of a series of international practical laboratories held for strategic guidance managing living cities: https://whc.unesco.org/en/news/2130.

The full list of events available at: https://whc.unesco.org/pg.cfm?cid=83&pattern=&id_keywords=839

have integrated the conservation of cultural heritage with urban development plans, policies, and processes using the HUL framework. Given that the concept of the HUL isn't solely applicable to historic centers designated as World Heritage sites, but rather extends its relevance to all historic centers, it's evident that there's still a considerable journey ahead.

With the internalization of debate on urban conservation, the need to reconcile conservation and development, a daunting challenge per se, was further complicated by the ambition of establishing standard-setting instruments for heritage conservation as the response to the challenges that affect historic centers worldwide. In this regard, the concept of the HUL stands as an exceptional example of a universally applicable methodology. Unlike some conventional approaches that risk homogenizing cultural aspects when applied beyond their original context, this approach, as recognized by Bandarin (2012), embraces local value systems as a pivotal factor in urban heritage conservation. By doing so, it transcends the confines of inflexible legal frameworks and propels itself into a realm of adaptable, context-sensitive methodologies that resonate with the diversity of urban landscapes worldwide.

The universality of the HUL approach holds another crucial aspect—its adaptability to existing legal systems. This is especially important for making the HUL methodology operational in contexts with rich national or local framework of management principles and conservation methodologies.

In recent years, the international community has assisted in the significant process of recognizing the value of World Heritage that has constituted an important opportunity for reevaluating management policies and programmatic tools within local administrations. Looking ahead, the international community's goal is to offer tangible solutions to the intricate challenges faced by historic cities. In this endeavor, a comprehensive understanding of the intricate interplay between heritage values, urban dynamics, and sustainable development serves as a versatile guide for policymakers, and urban and heritage practitioners worldwide.

References

Bandarin F (2012) From paradox to paradigm? Historic urban landscape as an urban conservation approach. Managing Cultural Landscapes, Routledge, London

Bandarin F, Van Oers R (2012) The Historic Urban Landscape: managing heritage in an urban century. Wiley-Blackwell, Oxford

Bini M, Capitanio C, Francini C (2015a) Firenze dal Centro alle Colline, Belvedere e percorsi panoramici. DIDA, Florence

Bini M, Capitanio C, Francini C (2015b) Buffer Zone: L'area di rispetto per il sito UNESCO Centro Storico di Firenze. DIDA, Florence

Cameron C (2008) Evolution of the application of outstanding universal value for cultural and natural heritage. In: What is OUV? Defining the outstanding universal value of cultural world heritage properties, vol XVI, pp 71–74. International Council on Monuments and Sites Monuments and Sites, Berlin

Council of Europe (1975) European charter of the architectural heritage. Council of Europe, Amsterdam

Francini C (2017) Da Centro Storico a Paesaggio Urbano Storico. Cultura Commestibile, n 209

Francini C, et al (2022) Conoscere, pianificare e ri-connettere i centri storici con il territorio. Historic Urban Landscape approach per il Centro Storico di Firenze. 'ANANKE 95, Milan

Francini C, Montacchini A (2021) Strategie post pandemia: l'approccio HUL (Historic Urban Landscape) per la gestione del Centro Storico di Firenze Patrimonio Mondiale UNESCO. Ri-Vista. Research for landscape architecture: Landscape design & COVID-19

ICOMOS (1964) The Venice charter: international charter for the conservation and restoration of monuments and sites. ICOMOS, Paris

ICOMOS (1975) The declaration of Amsterdam. ICOMOS, Paris

Jacobs J (1961) The death and life of great American cities. Random House, New York

Jokilehto J (2010) Reflection on HUL as a tool for urban conservation. Van Oers and Haraguchi

Lynch K (1960) The image of the city. MIT Press, Massachusetts

Mumford L (1938) The culture of cities. Secker and Warburg, London

Municipality of Florence (2019) World Heritage and relations with UNESCO Office. Firenze Patrimonio Mondiale. Appunti per un modello di valutazione di impatto sul patrimonio (HIA). curated by Francini C. Comune di Firenze, Florence

Municipality of Florence (2022) World Heritage and relations with UNESCO office. The Management Plan of the Historic Centre of Florence–UNESCO World Heritage site, curated by Francini C, Comune di Firenze, Florence

Riegl A (1982) Der moderne Denkmalkultus, sein Wesen und seine Entstehung. Published in English by Forster and Ghirardo The Modern Cult of Monuments: its character and its origins. Oppositions, p 25. Vienna

Rodwell D (2010) Historic urban landscapes: concept and management. Managing historic cities, p 99–104. World heritage series: papers, 27 [43]. UNESCO, Paris

Sitte C (1965) City planning according to artistic principles. Collins, London

UNESCO (1976) Recommendation concerning the Safeguarding and contemporary role of historic areas. UNESCO, Paris

UNESCO (2011) Recommendation on the Historic Urban Landscape, including a glossary of definitions. UNESCO, Paris

UNESCO World Heritage Centre (2005) Vienna memorandum on "World heritage and contemporary architecture-managing the historic urban landscape" and decision 29 COM 5D. UNESCO, Paris

WHC (2021) Call for action. Celebrating the 10th anniversary of the 2011 UNESCO recommendation on the Historic Urban Landscape. World Heritage Centre, Paris

WHC (2020) World heritage city lab. Summary report. World Heritage Centre, Paris

WHC (2023) Consolidated report on the third UNESCO member states consultation on the implementation of the 2011 recommendation on the Historic Urban Landscape. World Heritage Centre, Paris

Framing Planning and Conservation Activities in the Local Socio-Cultural Context: Ethnicity and Gender in Bamiyan

Mirella Loda and **Manfred Hinz**

Abstract This article illustrates the critical importance of research into the socio-cultural reality of Bamiyan (Afghanistan) for the cooperation projects aimed at preserving cultural heritage. The case study shows that exploring and ultimately deconstructing the categories of ethnicity and gender has helped overcome the impasse created by tensions and competition among different social and interest groups. As for the ethnic issue, a way out of this impasse might be found in putting ethnic self-definitions aside and searching for a reasonable compromise apropos the underlying economic interests. In regard to gender—a particularly sensitive issue in Islamic countries—the article questions the standardized way that urban planning addresses public spaces. It illustrates first that in the local context private dwellings de facto function as public spaces, especially for the female segment of the population, and therefore should be arranged more generously. Moreover, the article draws attention to the fact that the Islamic pilgrimage shrines (ziaratgah) play a crucial role for women, as the almost only public space (except the market) that they can attend without male company. It is, therefore of paramount importance to improve their quality as public spaces even if they are not included in the list of World Heritage assets.

Keywords Ethnicity · Gender · Urban planning in Islamic countries

M. Loda (✉)
Department of History, Archaeology, Geography, Fine and Performing Arts (SAGAS) and Laboratory for Social Geography (LaGeS), University of Florence, Via San Gallo 10, 50129 Firenze, Italy
e-mail: mirella.loda@unifi.it

M. Hinz
Philosophische Fakultät, Passau University, Passau, Germany
e-mail: manfred.hinz@uni-passau.de

M. Loda and P. Abenante (eds.), *Cultural Heritage and Development in Fragile Contexts*, Research for Development, https://doi.org/10.1007/978-3-031-54816-1_8

1 Premise

In-depth knowledge of the local socio-cultural context is crucially important in ensuring that cooperative activities achieve their desired outcome in the field of cultural heritage protection. This knowledge can help pinpoint solutions for safeguarding the heritage in a context of more general regional promotion and development goals. It can help mediate between international agencies' "universal" understanding of heritage and the local population's point of view, implicit in its social and cultural practices. What is more, it can help better understand differences existing in the local community with the aim of preventing dissatisfaction or even conflicts potentially arising as a result of the protective measures taken.

Appropriate consideration of the local socio-cultural context becomes particularly pressing when the heritage in need of protection is situated in an area like the Bamiyan Valley. Bamiyan is, on the one hand, characterized by the presence of historical structures, traditional dynamics, and ways of life and, on the other hand, subject to growing urban pressure, with increasing demand for modernization and development. A whole host of actors and different interest groups are at play, each carrying out its own actions and strategies aimed at taking advantage of the situation, while other segments of the local population remain voiceless.

In cases like this, measures to safeguard cultural heritage will be more effective if they are careful to take the long-term use of resources into consideration, combine interventions in favor of cultural heritage, development of the territory, and improvement of the local population's living conditions, and find a balance between the various social and economic interests at play.

We will use two examples to support these statements and show how, in the case of Bamiyan, socio-geographical analysis of the local context has helped to drive regional planning and/or heritage protection choices. The examples focus respectively on the role played locally by ethnicity and by gender.

2 Ethnicity

To correctly address the ethnic question in Bamiyan, we first need to understand the area's historical background and then to consider the issues surrounding its recent demographic growth.

2.1 Historical Background

Historically the Hazara population of Shi'a creed was the majority ethnic group inhabiting the valley[1] until the so-called "Hazara war" broke out in 1893/94, waged by King Abdur Rahman Khan,[2] during which over half of the Hazara population was wiped out, while the survivors were sold as slaves.[3] At the same time, a new Tajik population settled on the fertile valley floor and became the majority. From that point on, the ethnic question can basically be described as a tug of war between the two groups. During the Soviet occupation (1979–1989), Bamiyan again became the center of a quasi-state with a strong Hazara and Shi'a identity which was only overthrown in 1997 by the Taliban conquest (see Harpviken (1996), Grevemeyer (1988), Emadi (1997), Ibrahimi (2006)). Since 2001 and the overthrow of the Taliban, Hazara families from other parts of the province, from other provinces, and from abroad (above all refugees returning from Iran) have been drawing into the area. Finally, the return of the Taliban (August 2021) put an end, so to speak, to the direct competition between the Hazara and Tajik as a result of the emerging dominance of a third group, the Pashtu.

However, it should be stressed that words like Tajik and Hazara (but also Pashtu, etc.) are largely artificial umbrella terms that were coined and politically instrumentalized during the last few decades of civil war. They have, in some way, simplified the extremely varied ethnic mosaic that comprises Afghanistan, dividing it into three or four mega-groups which appear to exist in an ongoing state of tension among one another. In this process of roundup, the so-called "Tajiks" came to represent the Persian-speaking Sunni population, the Hazara the Persian-speaking Shi'a population, and the Pashtuns the Pashtu-speaking Sunni population. Furthermore, there are other groups who speak Turkic languages (Uzbeks, Turkmen, Qizilbash, etc.) both Sunni and Shi'a (Centlivres 1991; Schetter 2002; Barfield 2005; Dubow 2009; Siddique 2012). Some blame for the transferral of political and economic conflicts into artificial ethnic terms falls on the Afghan post-Bonn Constitution of 2004 which obliged all citizens to ascribe themselves to one ethnic group (Sahar 2014).

Consequently, the categories of ethnicity (as well as of gender) cannot be addressed in "substantialist" terms. This assumption is by now commonplace. Less self-evident, however, is the endeavor to re-translate supposedly ethnic (or for that matter gender) oppositions according to different interests and sometimes diverging daily practices. But, as we will see, only this "deconstruction" renders these oppositions manageable.

---

[1] For an historical and ethnographic overview over the Hazara populations in Central Asia, see Bacon (1951), Ferdinand (1959), Poladi (1989), Mousavi (1998), Uhrig (1999), Ibrahimi (2017).

[2] Abdur Rahman Khan sparked intense interest among the British public (Mahomed Khan 1900). In the same year, a tear-inducing literary account of the Hazara War was published (Hamilton 1900). This novel interestingly combines sympathy for the nationalistic enterprise of the Afghan king with sentimentality towards its Hazara victims.

[3] Slavery was only abolished in 1919 by King Amanullah Khan (Poladi 1989).

Table 1 Changing proportion of ethnic groups in Zargaran (*Source* authors' processing of LaGeS Household survey data 2017)

Year of settlement	Hazara	Tajik	Others	N
Up to 2010	78.3%	15%	6.7%	120
After 2010	89.9%	3.4%	6.7%	268
Tot				388

2.2 Demographic Growth and Segregation Patterns

As for demographic growth, between 1979 and 2017, Bamiyan's population rose from 7,355 to 51,856 inh. (LaGes 2018, p. 72), an increase of 600%. The biggest increase in population occurred after 2010.

This population explosion meant that Bamiyan quite suddenly boasted the fastest rate of urban growth in Afghanistan, posing new problems not only in terms of a sizeable increase in demand for housing and services (aspects typically factored into a strategic master plan), but also in terms of the balance between the different ethnic groups.

The number of the incoming Hazara population after 2001 has inverted the proportions of the two ethnic groups. While, in 1971, the proportion was 60% Tajik and 30% Hazara (Rasuli 1971, p. 29) and, in 2013, the Hazara grew to 83% and the Tajik group progressively decreased to 11% (LaGes 2018), pp. 72–76).

An exemplary case of this inversion is the village of Zargaran, where the Tajik accounted for 15% of settled households up to 2010 but only 3% after 2010 (Table 1).

The population distribution in the area is strongly segregated ethnically. By the end of the 1970s, using religious faith categories of belonging, Robert Canfield (Canfield 1973, 1986) described a segregated settlement structure, where the valley floor was occupied by Sunnis (corresponding to the Tajik ethnic group), the plateau by Imami (Shi'ites), and the eastern portion of the plateau by Ismaelites (Shi'ites) (Fig. 1).

The more recent arrivals not only confirm but also reinforce these divisions. For example, in Zargaran, one of the areas with the most rapid demographic growth (22% of its inhabitants have settled since 2019), the first area of expansion (until 2000) displayed a certain balance between ethnic groups. Today, however, the area is very much segregated (Fig. 2).

The Tajik population makes up just 7% of the total and is concentrated in the southwestern quarter of Zargaran, bordering the village of Dawoodi which is entirely Tajik.[4]

---

[4] This implies a tricky decision concerning the village perimeter to be adopted by the team of Florence University as part of the ongoing Zargaran regeneration project: either following an urban planning logic, which would place the village boundary at the end of the built-up area, or according to an ethnic-belonging criterion, which would place the Tajik group in the village of Dawoodi, to the west.

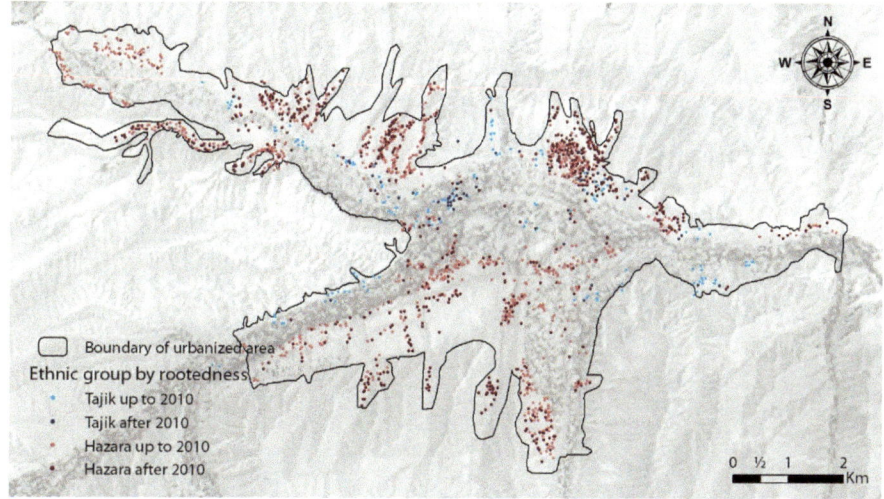

Fig. 1 Distribution of the Bamiyan population by ethnic group (2017) (*Source* LaGes (2018, p. 75))

2.3 Ethnicity and Cultural Heritage Protection

Against a historical background of repeated marginalization and persecution of the Hazara in Afghanistan, the reversal in the dominance of the two groups in the city of Bamiyan risks possible retaliation among the Hazara, and a priori resistance towards Hazara local government policies by the Tajik,[5] with serious political consequences, also regarding the central issue of cultural heritage protection.

A significant example can be found in the different reactions to the regional administration's decision to turn to UNESCO's guidance concerning what to do about the built-up areas on the valley floor, in front of the Western Buddha niche. This decision, aimed at protecting the cultural landscape, saw the Hazara population in favor, while the Tajik population, clearly opposed, supported among other things the reconstruction of the old bazaar in those very areas.

However, it was not so much a sensitivity towards protecting the heritage—although UNESCO's efforts certainly strengthened this feeling among the local community—that granted political feasibility to restrictive actions on the agricultural valley floor.

---

[5] Especially when we consider that the political management of Bamiyan province from 2002 to 2021 was strongly influenced by the Shi'a and Hazara party Hezb-e Wahdat Islami (Party of Islamic Unity). The varying degree of pro-Hazara orientation among the Bamiyan provincial governors after 2001 has been well-described in Adlparvar (2014). For the politico-ethnic self-definition of the Hazara, see Chiovenda (2016).

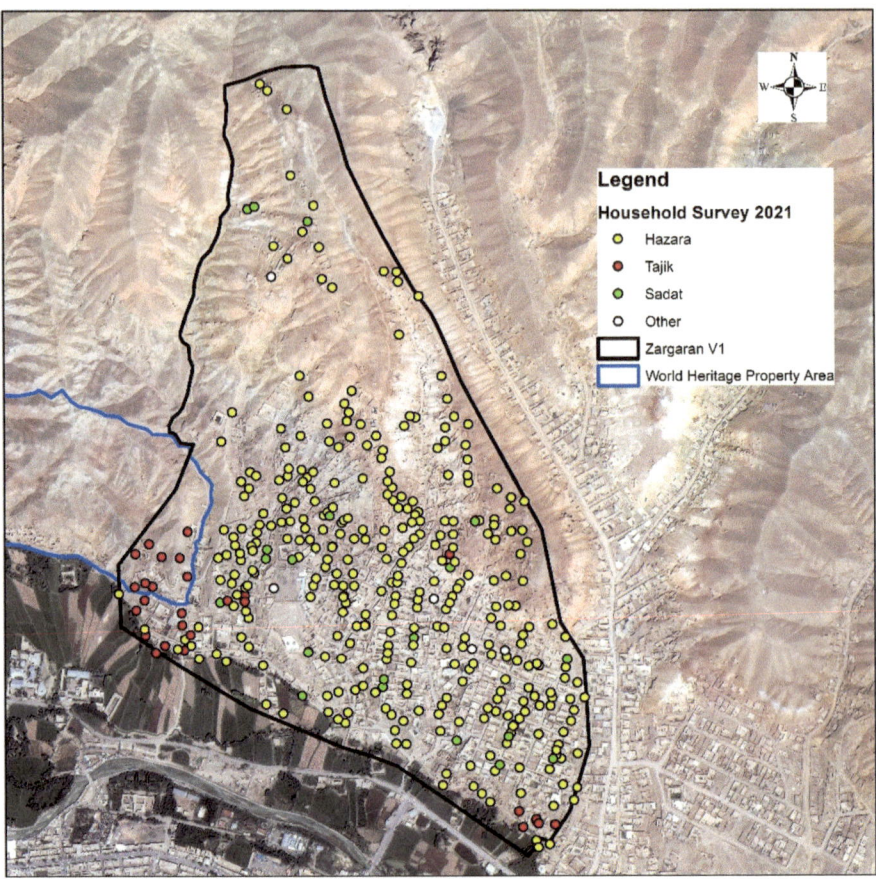

Fig. 2 Ethnic segregation in Zargaran (2021) (*Source* authors' processing of LaGeS household survey data 2021)

In the interplay between readiness to protect the landscape and economic interests, the ethnic dimension played a complex and misleading role,[6] making it harder to strike a balance between the different interest groups.

This became fully evident by combining and comparing the data collected in two direct surveys carried out while preparing the *Bamiyan Strategic Master-plan* (BSMP). The first survey consisted of a comprehensive analysis of the characteristics of the activities and of the 1,474 operators present in the new bazaar; the second survey consisted in mapping the agricultural lots in the buffer zones with informations about the owners (Fig. 3).[7]

---

[6] For a description of these vicissitudes see Adlparvar (2014, pp. 104 and 145ff).

[7] The following data was requested: name of the village, name and ethnic group of the owner, width of the parcel in jerib.

Fig. 3 Mapping of properties in the buffer zone (Detail of the LaGeS survey of agricultural lots 2018)

The surveys revealed an overwhelming presence of Hazara shops and workers in the rapidly expanding new bazaar, while the vast majority of the plots of land in the buffer zone belonged to Tajik families (Table 2).

As a consequence, the Hazara population favoured the actions implemented for the protection of the cultural landscape. The Hazara, whose economy is rooted in trading and in crafts (Monsutti 2004) and is dependent upon the new bazaar, would benefit greatly from the development of tourism, fuelled by the valley's inclusion on the world heritage list. On the contrary, the farmers living on the valley floor close to the UNESCO sites and the protected areas, mainly belonging to the Tajik, strongly opposed the same actions. For them, the restrictions, intended to protect the cultural landscape, were instead seen as an obstacle to economic activity and to improvements in living conditions.

In this context, it was inevitable that the cultural landscape debate would evolve into a politico-ethnical issue. The difficulties deriving from the different positions

Table 2 Ethnic belonging of the bazaar shop owners according to own definition (*Source* LaGes (2018, p. 101)	Ethnic group	N	%
	Hazara	1,111	75.4
	Sayed	174	11.8
	Tajik	119	8.1
	Ashar	26	1.8
	Ghizilbash	19	1.3
	Uzbek	19	1.3
	Pashtun	6	0.4
	Tot	1,474	100

on the topic of protecting the heritage were thus charged with tension, owing to the interference of ethnic factors but not caused by them.

The only way out of this endless process of clashing demands and mutual resentment would involve pushing the ethnic dimension aside once again. Documentation of the facts behind a great deal of the resistance against protecting the cultural landscape, while refraining from placing too much attention on ethnicity, might allow for a consideration of practical reasons or underlying interests, and to mediate between them (Adlparvar and Tadroz 2016). This has led to reflection not only on the dangerous side effects of an excessively restrictive protection policy but also on the "political" dimension of an apparently neutral, universal interest like the preservation of heritage and cultural landscape.[8]

3 Gender

The second example that we would like to give in order to stress the importance of preliminary socio-geographic research, concerns the necessity of a gender perspective, especially for cooperation projects in Islamic countries.[9]

Knowledge about mobility patterns, free-time activities, and the symbolic meanings given to places by the local population can function as a starting point for understanding the socio-cultural practices of the local community, especially in regard to the female population, often voiceless, and for consistently addressing the cooperation measures.[10]

3.1 A Limited Mobility

According to the data collected through a Household survey in 2017, about 64,600 trips are made on an average weekday in the city of Bamiyan (LaGes 2018, p. 138) and the number of people commuting daily is 92% of the total. However, the distribution of trips varies significantly by gender and age (Fig. 4). This distribution first of all reflects the demographic structure of the city, 63.6% of whose population is under 25 years old and only 5% which is over 60 years old (LaGes 2018: p. 72): indeed, the most active age group is that between 15 and 20 years old and 67% of the trips involve people between the ages of 10 and 25. On the contrary, only 14% of the trips involve people over 40 years old.

---

[8] Within the framework of heritage studies, a recent volume reflects upon heritage managment as a means of "reconcilition" (Ringbeck 2022, pp. 439–444).

[9] An ethnological analysis of the situation of women in central Afghanistan is offered in Tapper (1991), the "classic" text remains (Dupree 1990).

[10] For gender relations in the employment sector before the Taliban's seizure of power see Islamic Republic of Afghanistan (2019); for the present situation of the female population see UN-Women (2022).

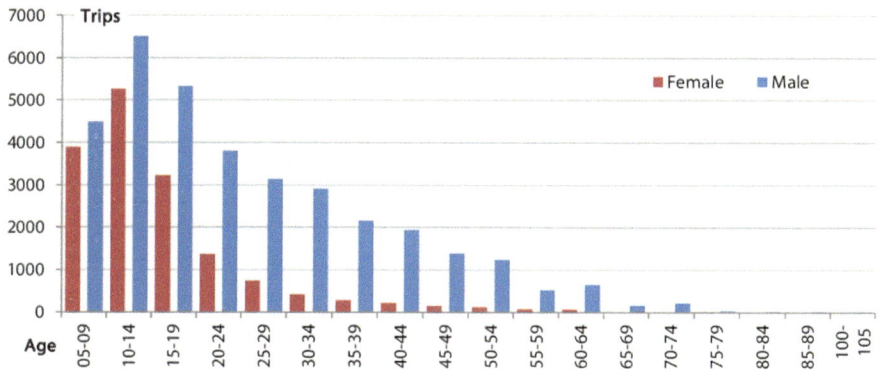

Fig. 4 Distribution of daily trips of Bamiyan residents according to age and gender (*Source* LaGes (2018, p. 139))

However, the figure clearly indicates that differences in mobility behaviors are also gendered. Only 32% of the total daily trips are made by females, compared to 68% of trips which are made by men. In addition, the age distribution of women's mobility is even more strongly concentrated among the very young population, with most trips made between the ages of 10 and 15, and a sharp decline after the age of 20. This decline is such that 86% of women's trips are made under the age of 25, before the end of school-going age.

Moreover, in a context where more than 90% of travel is due to school attendance (58%) or work (38%), the distribution of female travel between school and work is 78% and 19%, respectively, while for males the percentages are 48 and 48% (Fig. 5).

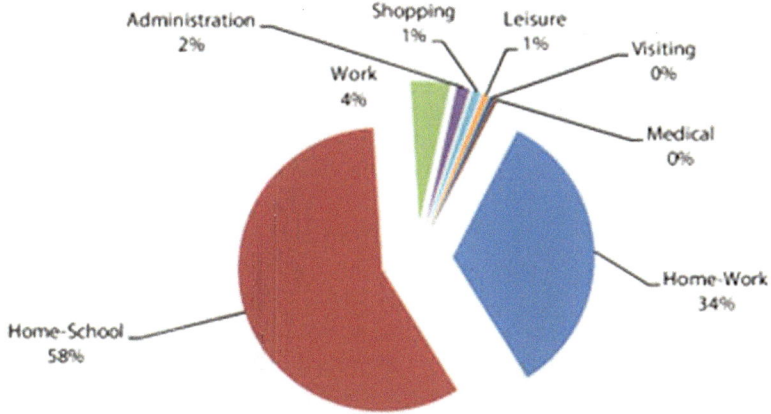

Fig. 5 Distribution of daily trips of Bamiyan residents according to reason (*Source* LaGes (2018, p. 140))

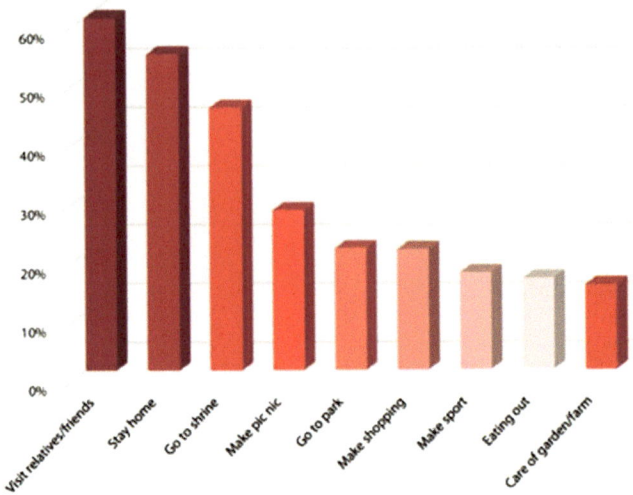

Fig. 6 Favourite leisure time activity by gender (N = 2,030) (*Source* LaGes (2018, p. 82))

3.2 Social Practices and Public Space

This data highlights how domestic barriers confine the fundamental sphere of movement among the adult female population, whose interaction with the rest of the urban context is already extremely limited.

The same considerations apply when looking at activities carried out during leisure time or for socialization purposes.

The social life of the community is predominantly structured around interaction with networks of relatives and friends, while the home acts as both an active and passive destination when it comes to visits to/from relatives and friends, the preferred free-time activity of the majority (more than 2/3) of those interviewed (Fig. 6).

As such, it is significant that as many as 20% of the interviewees indicated their place of residence among their three favorite places, and that typically residential areas, villages such as Zargaran, Mullah Gholam, Sayed Abad or Haidar Abad, appear among the most frequently cited places (Fig. 7).[11]

However, even from this point of view, substantial gender differences emerge. The home stands as the preferred space for over 60% of the women in their free time, while the percentage is significantly lower for the male population, where interviewees also mention outdoor or sports activities (Fig. 8).

Therefore, it is worth considering on the one hand the social function (broadly speaking) played by homely spaces as tendentially public spaces[12]; on the other hand,

---

[11] References to the heritage protected by UNESCO are limited to the Buddhas and Shar-e Zohak, mentioned by 448 and 169 interviewees, respectively. The low symbolic value of Bamiyan's UNESCO properties for the local population is also discussed in Wyndham (2015, p. 392).

[12] Considering the extension of the local family networks.

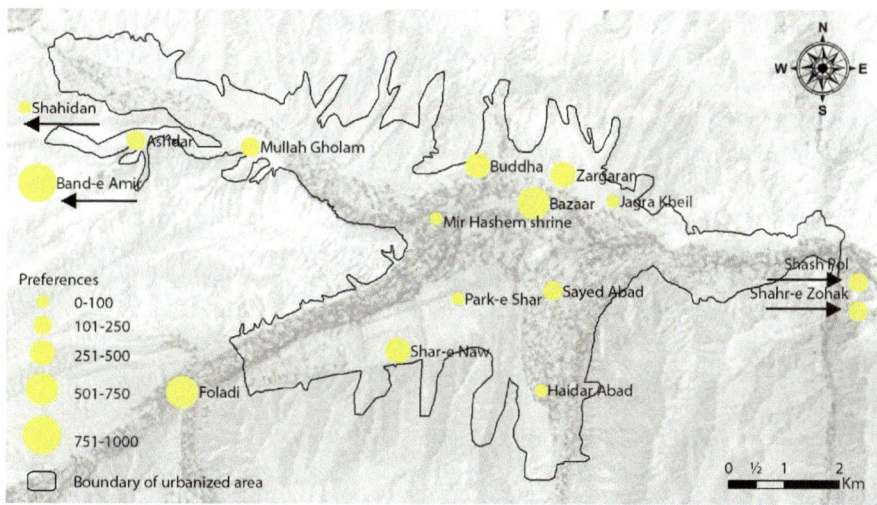

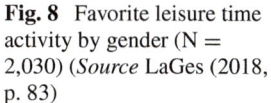

Fig. 7 Most frequently cited favorite places (*Source* LaGes (2018, p. 78))

Fig. 8 Favorite leisure time activity by gender (N = 2,030) (*Source* LaGes (2018, p. 83)

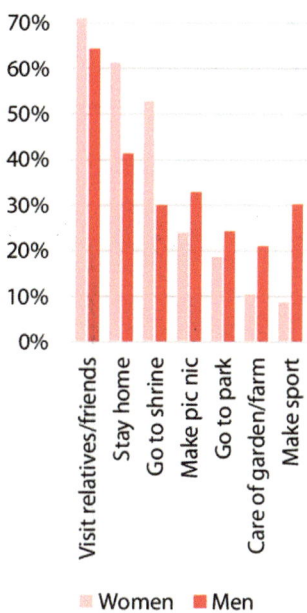

it is worth deepening what might be the appropriate conceptualization of open spaces which qualify as public spaces in the local urban context, as well as the dimensions and the material assets which would be appropriate.

Help with this might again be sought from the results of socio-geographic research, especially from the data concerning perceptions and symbolic meanings bestowed on different places by the local population.

Interestingly, shrines emerge as important destinations for free-time activities among members of the female population, not only for religious reasons but also above all as public spaces, typically accessible on foot and without the need for male accompaniment.

Therefore, alongside the famous Mir Hashem shrine (Ziaratgah), also mentioned by the male interviewees, albeit to a lesser extent, the spectrum of favorite places quoted by the interviewees also includes numerous smaller shrines (such as Mir Said Ali Jakhsuz or the Khwaja Sabz Push shrine of fertility), which are used as places of worship and, at the same time, as neighborhood meeting places by the female population (Fig. 9).

The role of gender as a discriminating factor in the organization of moments of socialization and public social life is reflected in the comparison between genders when other factors of differentiation are the same (age, income). This can be seen, for example, when isolating the under-25 s with a high income.

An analysis of the services that people lack confirms this reading (Fig. 10).

Services which the inhabitants would like to see strengthened function as a significant indicator both of the shortcomings of the context under examination and their

Fig. 9 Mir Hashem shrine, one of the favorite shrines mentioned by female population (*Photo* M. Loda 2023)

Fig. 10 Missing facilities (*Source* authors' processing of LaGeS Household survey data 2017)

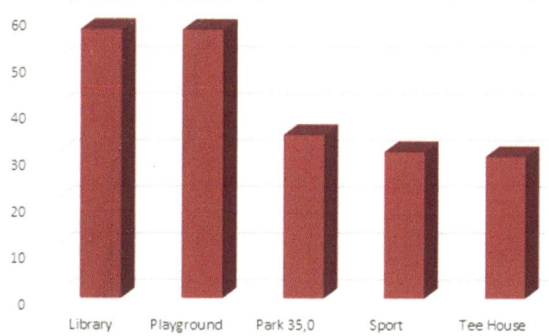

model of reference (actual or ideal). According to the sample interviewed, libraries and playgrounds, followed by parks, sports facilities, and tea houses, are among the amenities that people lack the most in their neighborhood, and that they feel would significantly improve their quality of life.

As already observed in other urban contexts in the country (LaGes 2013), interest in libraries not only reflects the great prestige traditionally ascribed to literature and "knowledge" in general in Afghan culture but also reflects an appreciation for protected public spaces, which the female population can attend by themselves.

As expected, opinions regarding facilities that are lacking vary according to gender. Among women, there is a higher percentage who complain of a lack of playgrounds. Nevertheless, in this case, the gender differences are not noteworthy. Quite a weak tendency can be deduced on the part of females in outlining their specific needs, also in comparison with what can be seen in other Afghan towns and cities.[13] This is another reason for deepening our understanding of the needs of this segment of the population while preparing urban intervention.

Awareness of the role played by the home as a space of socialization (as well as of women's work in the case of 10% of families in Zargaran) helps drive urban planning choices. Alongside this, for the specific needs of women, it is advisable from a strategic point of view to recognize and promote the role of libraries and shrines both as public spaces and as places of local cultural heritage.

Therefore, interventions to protect and promote shrines and the surrounding areas, perhaps backed by international donors, would be an important gesture and would help narrow the gap frequently found—and criticized in international literature—between what the international bodies consider cultural heritage and the cultural heritage that is part and parcel of local customs.

---

[13] The female population more clearly separates both needs typically linked to the maternal role (greater interest in playgrounds) and preferences connected to the modern evolution of lifestyles, albeit in the context of a local society that tends to organize itself in a homo-social way: therefore, there is even greater interest in sports facilities than among the male component, but only if strictly organized according to a principle of gender segregation (LaGes 2013: p. 64).

4 Conclusions

A precise socio-cultural analysis, both before and during the implementation of cooperation projects in the field of urban planning and management, as well as of cultural heritage protection, has proved crucial for securing quality results. The case of Bamiyan demonstrates that—even when it comes to an especially sensitive issue like ethnicity—it is still possible to find appropriate solutions for the protection of cultural heritage and to relieve social tensions. As for the gender question, the case of Bamiyan confirms that special attention to the socio-cultural context is particularly useful for urban planning in Islamic countries, going beyond abstract approaches and developing autonomous concepts and schemes that fit into the local way of life.

References

Adlparvar N (2014) When glass breaks, it becomes sharper. De-Constructing ethnicity in the Bamiyan Valley, Afghanistan. PhD thesis, University of Sussex, Institute of Development Studies, http://sro.sussex.ac.uk. Last Accessed 21 July 2020

Adlparvar N, Tadroz M (2016) The evolution of ethnicity theory: intersectionality. Geopolit Dev Inst Dev Stud (IDS) Bull 47(2):123–136. https://www.researchgate.net/publication/301903 705_The_Evolution_of_Ethnicity_Theory_Intersectionality_Geopolitics_and_Development/ link/57445a2908ae298602f40c88/download. Last Accessed 26 Apr 2020

Bacon E (1951) Inquiry into the history of Hazara Mongols of Afghanistan. Southwest J Anthropol 7:230–247

Barfield T (2005) Afghanistan is not the Balkans: ethnicity and its political consequences from a central eurasian perspective. Cent Eurasian Stud Rev 4(1):4–9

Canfield R (1973) Faction and conversion in a plural society. Religious alignments in the Hindu Kush, University of Michigan Museum of Anthropology, Ann Arbor

Canfield R (1986) Ethnic, regional, and sectarian alignments in Afghanistan. In: Bonazizi A, Weiner M (eds) The state, religion, and Islamic politics: Afghanistan, Iran, and Pakistan, pp 75–103. Syracuse UP

Centlivres P (1991) Exil, relations interethniques et identité dans la crise afghan, Revue du monde musulman et de la Méditerranée, pp 70–82, https://doi.org/10.3406/remmm.1991.1492

Chiovenda MSK (2016) Cultural trauma, history making, and the politics of ethnic identity among Afghan Hazaras. PhD thesis, University of Connecticut Graduate School, https://digitalco mmons.lib.uconn.edu/cgi/viewcontent.cgi?article=7545&context=dissertations. Last Accessed 03 Jan 2019

Dubow B (2009) Ethnicity, space, and politics in Afghanistan. University of Pennsylvania, Urban Studies Program, papers 13, https://repository.upenn.edu/senior_seminar/13. Last Accessed 12 Apr 2019

Dupree NH (1990) The women of Afghanistan. Kabul, http://afghandata.org:8080/xmlui/handle/ azu/3791. Last Accessed 12 Jan 2016

Emadi H (1997) The Hazara and their role in the progress of political transformation in Afghanistan. Cent Asian Surv 16:363–387

Ferdinand K (1959) Preliminary notes on Hazara culture. Ejnar Munksgaard, København

Grevemeyer J-H (1988) Ethnicity and national liberation: the Afghan Hazara between resistance and civil war, Le fait ethnique en Iran et en Afghanistan, pp 211–218. Editions du CNRS, Paris

Hamilton L (1900) A Vizier's daughter. a tale from the Hazara war. Murray, London

Harpviken KB (1996) Political mobilization among the Hazara of Afghanistan 1978–1992. Institutt for Sosiologi, Oslo

Ibrahimi N (2006) The failure of a Clerical Proto-State: Hazarajat 1979–1984, Crisis States Research Centre, Working paper 6, London

Ibrahimi N (2017) The Hazaras and the Afghan state. Rebellion, exclusion and the struggle for recognition. Hurst, London.

Islamic Republic of Afghanistan (2019) National statistics and information agency, women and men in Afghanistan 2018. Kabul. https://www.CSO,%20Women-and-men-in-Afghanistan-2018_f ull.pdf. Last Accessed 02 May 2020

LaGes (Laboratorio di Geografia Sociale, Università degli Studi di Firenze) (2013) Herat strategic master plan. A vision for the future. Polistampa, Firenze

LaGes (Laboratorio di Geografia Sociale, Università degli Studi di Firenze) (2018) Bamiyan strategic master plan. Polistampa, Firenze

Mahomed Khan S (1900) The life of Abdur Rahman Amir of Afghanistan, vol 2. Murray, London

Monsutti A (2004) Guerres et migrations. Réseaux sociaux et stratégies économiques des Hazaras d'Afghanistan. Institut d'Ethnologie, Neuchâtel-Paris

Mousavi SA (1998) The Hazaras of Afghanistan. A historical, cultural, economic and political study. Curzon, London

Poladi H (1989) The Hazaras. Mughal Publ, Stockton

Rasuli GO (1971) The economy of Bamiyan. Economic development and population growth of Bamiyan, Afghanistan. Kabul University, Kabul

Ringbeck B (2022) World heritage and reconciliation. In: Albert M-T, Bernecker R, Cave C, Prodan AC, Ripp M (eds) 50 Years word heritage convention: shared responsibility–conflict & reconciliation, pp 439–444. Springer, Federal Foreign Office, Germany. https://doi.org/10.1007/978-3-031-05660-4

Sahar A (2014) Ethnicizing masses in post-bonn Afghanistan: the case of the 2004 and 2009 presidential elections. Asian J Polit Sci 22(3):289–314. https://doi.org/10.1080/02185377.2014.945941

Schetter C (2002) Der Afghanistankrieg-Die Ethnisierung eines Konflikts. Internationales Asienforum 33(1–2):15–29

Siddique A (2012) Afghanistan's ethnic divides. CIDOB Policy Research Project, Oslo

Tapper N (1991) Bartered brides. Politics, gender and marriage in an Afghan tribal society. Cambridge UP

Uhrig R (1999) Die Ethnie der Hazara in Afghanistan. Internationales Asienforum 30:27–46

UN-Women (2022) Women's rights in Afghanistan one year after the Taliban takeover, https://www.unwomen.org/en/news-stories/in-focus/2022/08/in-focus-women-in-afghan istan-one-year-after-the-taliban-takeover. Last Accessed 01 Nov 2022

Wyndham C (2016) Investigating values ascribed to cultural heritage sites in Bamiyan by residents of the Bamiyan Valley (May 2015). In: Emmerling E, Petzet M (eds) The giant Buddhas of Bamiyan II, safeguarding the remains (ICOMOS XXI), pp 389–400. Baessler Verlag, Muenchen

Improving Urban Quality Through Land Titling? Considerations from the Bamiyan Case

Mirella Loda⊙**, Bashir Amiri, and Nipesh Palat Narayanan**⊙

Abstract The article discusses the Afghan land titling policies based on the case of Bamiyan Valley. It first presents the terms and conditions of the land titling policy in the Islamic Republic of Afghanistan since 2017, and then it illustrates its impact on the informal settlement of Zargaran (Bamiyan) based on the results of two surveys conducted in 2017 and 2021. As is the case elsewhere around the world, the assignment of formal property titles is generally welcomed by the majority of the population. Moreover, doing so has proven to encourage investments in the improvement of private establishments, and even in facilities for community purposes, thanks to the remarkable social bond that exists between the settlers. However, the denial of the entitlement in the parcels of Zargaran located inside and next to the UNESCO buffer zone has prevented the titling policy from reaching its full potential in terms of improvement of the social fabric and urban quality. Moreover, the increase in real estate values observed in the area calls for social policy measures to accompany the titling policy, so as to avoid the eviction of poorer segments of the population.

Keywords Land titling · Bamiyan world heritage site · Real estate market

M. Loda (✉) · B. Amiri
Department of History, Archaeology, Geography, Fine and Performing Arts (SAGAS) and Laboratory for Social Geography (LaGeS), University of Florence, Via San Gallo 10, 50129 Firenze, Italy
e-mail: mirella.loda@unifi.it

B. Amiri
e-mail: bashir.amiri@stud.unifi.it

N. P. Narayanan
Centre Urbanisation Culture Société, Institut National de La Recherche Scientifique (INRS), Québec City, Canada
e-mail: nipesh.palat.narayanan@inrs.ca

© The Author(s) 2024
M. Loda and P. Abenante (eds.), *Cultural Heritage and Development in Fragile Contexts*, Research for Development, https://doi.org/10.1007/978-3-031-54816-1_9

1 Introduction

Land titling has long been a contested topic, both in policy and academic circles, and has occupied a vast part of the debate on the possibility—or even the opportunity[1]—to manage the widespread processes of intense and spontaneous urban growth taking place in the Global South.

Tenure security at a household scale and liveable spaces along with social amenities at a neighbourhood scale are crucial ingredients for a decent standard of living in the modern urban world. However, land is a complex entity, which is ideologically governed (Ghertner 2020; Ward 2021), politically contested (Balakrishnan and Pani 2021; Wani 2021) and socially constructed (Anwar 2012; Opoko et al. 2020). Therefore, the literature on land titling (and allied topics around land) is largely heterogeneous.

The land titling debate initially began as a way of alleviating the poverty that plagued inhabitants of rapidly expanding settlements in urban areas of the Global South.

Transferring the informal capital represented by these settlements, which Hernando de Soto in 2001 calculated at 9,3 Trillion US Dollars worldwide (Soto 2001, pp. 29f) into the formal market, would alleviate, if not resolve, the struggles of the urban poor.[2] This assumption largely inspired land titling policies that have been carried out by the World Bank and by many international NGOs since the 1990s.[3]

However, strong criticisms arose against land titling policies. An overcritical view of (legal) formalization as an exogenous form of statal control and normalization unfolded into a somehow romanticized view of informal settlements as self-governed, constitutively disobedient to any legal regularization and inherently 'democratic'.[4]

---

[1] Some authors, foremost Anna Roy, see urban planning as a device of social control and informality as intrinsically subversive: 'Urban informality makes possible an understanding of how the slum is produced through the governmental administration of a population. (…) In this sense, urban informality is a heuristic device that serves to deconstruct the very basis of state legitimacy and its various instruments: maps, surveys, property, zoning, and, most importantly, the law' (Roy 2012, p. 132). The thesis, that it is the formal state apparatus which produces informality is already put forward by Roy (2005). For the numerous followers of Roy's approach see Simone (2001), Watson (2009), Porter (2011) e.a. Roy's theses are applied into Afghanistan in Calogero and Schütte (2018, pp. 14, 15).

[2] For positions akin to de Soto see Durand-Lasserve (2002), Arnott (2008), Fernándes (2011), Riley and Wakely (2011).

[3] For a description of this alignment see Davis (2006, pp. 71ff). These policies were, however, not valued uncritically within the World Bank itself, see Dowell and Clarke (1996).

[4] The founder of this tendency is Turner (1967, 1983); see also Kellett and Napier (1995). Turner's line of thinking is today continued in Lizarralde (2011). Others Bredenoord and Lindert (2010) opt for an 'assisted self-help housing policy'. The romanticizing tendency is observable in Bayat (2000), Koster and Nuijten (2016), Lutzoni (2016), Schindler (2017), Kucina (2018), Mantia (2018).

Criticisms were also directed at the supposed effectiveness of land titling in alleviating poverty.[5] Finally, some authors observed that land titling policies do not necessarily engender urban upgrading projects and that the latter are much more helpful in alleviating poverty than the former (Mukhija 2001; Majale 2008; Das and Takahashi 2009; Mistro and Hensher 2012; Dovey 2012; Russ and Takahashi 2013; Devkar et al. 2019; Heikkila and Harten 2019). As we will see in the following pages, our results challenge this question, introducing an urban argument in individual property rights debates.

For the purposes of this paper, we conceptually subdivide the vast theoretical discussion on land titling into two thematic categories.[6]

First are those authors who investigate the topic from an economic perspective, debating whether land titling policies can be considered effective tools to reduce poverty.[7]

Land tenure/titling as a means to poverty alleviation and its critique (Soto 2001; Payne et al. 2009).

Impacts of land titling on human development (Abdulai et al. 2011; Ali et al. 2014; Janvry et al. 2015; Field 2007; Galiani and Schargrodsky 2010).

Nuances of the land titling process itself (Jonnalagadda et al. 2021; Meinzen-Dick and Mwangi 2009; Toulmin 2009).

Second are those authors who understand land titling along with urbanization, and investigate it within the larger urban context.

Understanding larger urban issues due to (and via) land titling processes (Cheng et al. 2019; Deininger et al. 2014; Goldstein and Udry 2008; Ho and Spoor 2006).

Understanding urbanization and land as entangled (along with its impact on socioeconomic development (AlSayyad 2004; Benjamin 2008; Gilbert 2007; McFarlane 2020; Upadhya and Rao 2022).

In this paper, we will for the most part remain within the second field of study. Using primary data from Bamiyan (Afghanistan)[8] we will move towards a nuanced understanding of impacts and future pathways for land titling processes. We are interested in analysing to what extent land titling can be conceptualized in order to produce multiple outcomes; as a potential tool for improving both private housing and urban quality. Focusing on the neighbourhood rather than on the single household,

---

[5] See Bromley (1990), Gilbert (2002), Jones (2017). Mike Davis calls De Soto 'the Guru of neoliberal populism' (Davis 2006, p. 79).

[6] For a concise but thorough overview see recently (Boanada-Fuchs and Boanada-Fuchs 2018).

[7] It should be kept in mind that the concept of 'informality' firstly emerged within the *International Labour Organization* (ILO) in an attempt to grasp new, emerging forms of labour (Bangasser 2000; Benanav 2019).

[8] The data have been collected in 2017 and in 2021 by the team of LaGeS (Laboratory for Social Geography) at Florence University within the framework of two research and cooperation projects with the Ministry of Urban Development and Land funded by the Italian Agency for Development Cooperation (AICS). The first project resulted in the preparation of the *Bamiyan Strategic Master Plan* (2018) (LaGeS 2018). The second project (ongoing) aims at improving the urban quality in Zargaran, the fastest growing neighbourhood in Bamiyan.

we will reflect on the extent to which tenure security could be mobilized to improve land use and to carry out participatory urban upgrading projects.

The paper is divided into four parts. The first section frames Afghan land titling policies in the context of the intense urbanization process that the country has experienced in the last two decades, in particular the Bamiyan area. The second part re-approaches key points in the land titling debate in light of the primary data related to Bamiyan. The third part addresses land titling as a potential tool for improving housing and urban quality. The fourth part discusses problems connected to the implementation of land titling policies in the cultural heritage protection context. The article closes with a series of concluding remarks.

2   Urban Growth in Afghanistan and in Bamiyan

In Afghanistan, land titling policies took concrete shape with the passing of the 2017 law 'RRUIP—Regulation on Registration of Urban Informal Properties'.[9] This law was an attempt to tackle the massive urban migration taking place throughout the country, one of the fastest urbanization rates in the world.

UN-Habitat stated in 2017, the year of the implementation of the land titling campaign: 'Afghanistan is still a predominately rural society with an estimated 76% of the population living in rural areas. However, this situation is rapidly changing. Afghan cities are growing at an estimated rate of around 4% per year, one of the highest in the world; and the urban population is expected to continue to grow at an average of 3.14% up to 2050. In 1950, only one out of every 20 Afghans lived in cities. In 2015, 8.5 million or one out of every four (27%) Afghans lived in cities; and by 2060 one in two—50% of the population—will live in cities' (UN-Habitat 2017, p. 6). This pace of urbanization led to an unrestrained expansion of new 'informal

---

[9] The Afghan land titling policy was implemented since 2017 in selected Afghan municipalities (among which Bamiyan) within the broader framework of the Afghanistan Land Administration System Project (ALASP) funded by the World Bank which comprised also a digital cadaster system and other items. For an assessment report 2019 see Islamic Republic of Afghanistan (2019).

settlements'[10] which currently provide housing for around 85% of the population, and often fall dramatically below basic urban planning standards.[11]

Within the framework of this urban growth, Bamiyan is one of the cities in Afghanistan undergoing the most expansion.

The total population living in the Bamiyan urban area was estimated in 2017 at about 51,852 inhabitants[12]—an incredible increase compared to the 7,355 inhabitants recorded in the previous census of 1979. The demographic growth began after the fall of the first Taliban regime (2001) fostered both by the return of refugees and internally displaced people (IDP) as well as by the urban drift of the rural population from other districts in the province. But this growth has gathered momentum in recent years. Compared to the UN-Habitat SoAC data from 2014 (Amiri and Lukumwena 2018), the LaGeS survey of 2017 highlights a demographic increase of no less than 33.0% in three years; a population increase that continued growing at the same rate until 2021, according to data collected in April of that year in the neighbourhood of Zargaran.[13]

Given the great variability in the orographic subdivisions of the valley, the population growth and expansion of the urban fabric have mainly been concentrated in two areas located on the south-facing mountain slopes alongside the valley, where the population density is much higher than the Bamiyan average.

One of these areas is Zargaran, the neighbourhood which we will concentrate on in the following paragraphs, in order to analyse the impacts of the Afghan land titling policy (Fig. 1).

---

[10] Informal settlements are defined by UN-Habitat (2015a, p. 1) as 'residential areas where (1) inhabitants have no security of tenure vis-à-vis the land or dwellings they inhabit, with modalities ranging from squatting to informal rental housing, (2) the neighbourhoods usually lack, or are cut off from, basic services and city infrastructure, (3) the housing may not comply with current planning and building regulations and is often situated in geographically and environmentally hazardous areas'. In practice, however, and depending on the geographical reality each time, we find situations that greatly differ from one another, as much from the point of view of the legal and urban regime, as from that of material conditions. Indicative of this is the multiplicity and semantic variety of terms associated with the concept: illegal, unplanned, marginal, squatter and so on. Although informality is complex and informal settlements is a contested term, we use it as a heuristic category. We do not mobilize a universal understanding, but, for this chapter 'informal settlement' is a category that the local municipality/state uses. Although at times contradictory and not comprehensive, this local understanding of settlement categorisation is important for our analysis.

[11] These data vary slightly in the literature. UN-Habitat 2017 gives nationwide 70% of Afghan houses in informal settlements, for the Kabul area 80% (UN-Habitat 2017, pp. 52, 56). Amiri, Lukumwena, give 82% of houses in informal settlements in 2018 for the Kabul area (Amiri and Lukumwena 2018, pp. 348f). The rate of poor housing in Afghanistan is even higher than that of informal housing: 'The majority (86%) of the current urban housing stock in Afghanistan can be classified as 'slum' based on the UN-Habitat definition of not fulfilling one or more of the following criteria: (i) security of tenure (ii) access to a safe water source, (iii) improved sanitation; (iv) durable, structurally sound housing materials; and (v) adequate living space' (UN-Habitat 2017, p. 34).

[12] The estimation is based on the results of the household survey carried out by LaGeS in the period of April–May 2017; the survey involved over 2,000 households (LaGes 2018).

[13] The data was collected through the household survey carried out by LaGeS in April–May 2021 within the framework of the second cooperation project (see note 8).

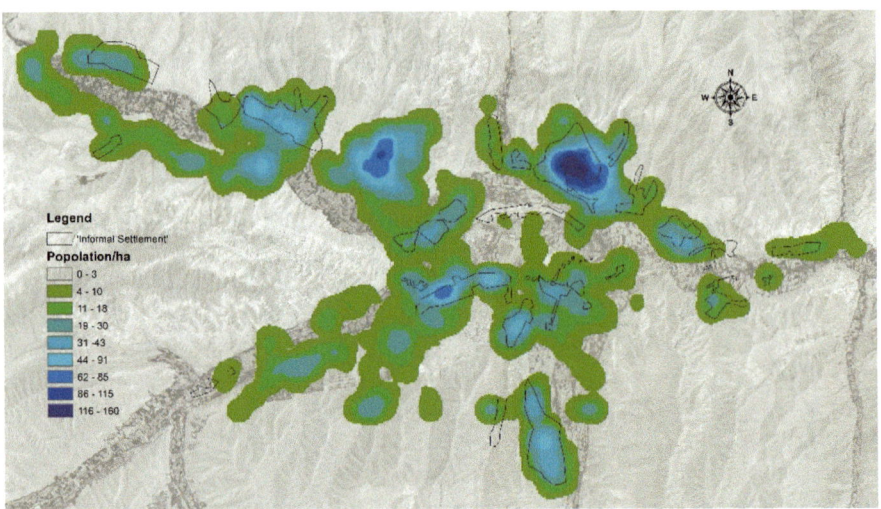

Fig. 1 Population density in Bamiyan 2017 (*Source* LaGes (2018))

3 Key Issues with Land Titling

Among the several issues relating to informal settlement expansion on a global scale, two occupy a prominent place in the context discussed here: 1. The understanding of settlement processes as the result of competitive dynamics; 2. The contextual nature of land security forms. In this section we address these issues in light of the primary data collected in Zargaran in the spring of 2021.

3.1 Occupancy Urbanism

Until 2003, Zargaran consisted of just a few farmhouses built on the hill above the highest irrigation channel, to preserve the fertile soil on the valley floor. But the neighbourhood has expanded very quickly since 2003 when the municipality started distributing plots of land to migrants (returnees and IDPs). Between 2003 and 2010, the municipality distributed a total of 224 plots in the south-eastern corner of Zargaran. At the same time, however, many people started to settle in other parts of the neighbourhood (UN-Habitat 2015b).

Zargaran's population, estimated at 6,604 inhabitants in May 2017 (first LaGeS household survey (LaGes 2018), grew to 8,866 inh. in 2021 (second LaGeS household survey). In only 4 years, there was a 34% increase in the population, with 22% of the households settling between 2019 and 2021. The built-up area also continued to expand at a rapid pace. There were 633 buildings captured by satellite in 2015 (UN-Habitat (Ahmadi 2018)), 1,050 in June 2017 (LaGes 2018) and 1,437 in June

Fig. 2 Buildings on steep slopes (*Photo* M. Loda 2019)

2021 according to the UN-Habitat survey conducted for the implementation of the land titling campaign. Of these buildings, according to the UN-Habitat classification, 1,310 were residential and the rest were businesses or mixed-purpose (residential and business).

The settlement has continued to expand in line with a settlement model, which traditionally preserves the fertile soil of the valley floor; the new buildings, however, were being built up mountain slopes with gradients of up to 29%, which poses serious problems for the provision of services (Fig. 2).

It is unclear to what extent the expansion of Zargaran (as well as other 'informal settlements' that sprang up in the valley during the same period) may have been influenced by illegally acting brokers.[14]

---

[14] Based on the information gathered on site, unplanned settlement expansion has been the result of the illegal sale of state land by certain brokers (belonging to both the Hazara and the Tajik), in a context of ever-weakening land control by the central authorities and at times even the local government's complicity. The phenomenon has been observed in many other urban areas in Afghanistan. Habib (2011, p. 369) uses the term 'commanders', ethnically affiliated, for these real estate developers in the Kabul metropolitan area. UN-Habitat (2017, p. 25) cautiously speaks of 'private developers' who managed and (in a certain sense) 'planned' the informal settlements around the country. Already De Soto registered the 'omnipresence of Illegal Real Estate Brokers' (Soto 2001, p. 186), while Davis (2006, pp. 41 and 82) speaks of 'control by powerful locals'. Clark (2023) has recently analysed for the *Afghan Analysts Network* the decrees of the *Amir ul-mo'emin*, Haibatullah Akhundzada, from 2016 to 2023 (Clark 2023). Out of the 65 issued decrees, 6, all issued after the instalment of the Emirate in August 2021, regard the problem of land grabbing. In order to put an end to land distribution by powerful locals, the Amir has bound all such processes to his personal permission and, thus, effectively monopolized new distributions of land.

However, it does not seem inconsequential that this process took place in a context characterized by a high degree of 'ethnic' acceptance regarding the new Hazara population by the local government,[15] and a support policy for the Hazara constituent, pursued by that government in the years prior.[16] Such agreement is reminiscent of 'vote bank' politics, albeit in a radically different social, political and economic context from that of Bangalore, where the phenomenon was initially analysed (Benjamin 2008). In Zargaran the political exchange would aim to achieve not so much infrastructural neighbourhood improvements, but rather substantial governmental support for changes in land use, in the context of an ethnically based negotiation between the new Hazara population seeking residential space, and the inhabitants of the neighbouring Tajik village of Dawoodi, for whom Zargaran land had previously signified space for their agro-pastoral practices.[17]

In fact, the adjudication of legal property titles can, per definition, only recognize individual property rights, which implies the tacit expropriation of all forms of customary, 'traditional', communal property claims.[18]

Zargaran's excessive expansion thus confirms the negotiating/conflictual nature of the urbanization process echoed by the concept of 'occupancy urbanism': 'Land (rather than Economy) as a conceptual entry, helps reveal subtle, often stealth-like and quiet, but extensive forms of political consciousness. This perspective avoids a conceptual 'prison house' built around assumptions of a predestined development trajectory, or the constraint of uneven terrains viewed as fractures, and relationships ordered within a taxonomy. 'Occupancy urbanism' instead views cities as consisting of multiple, contested territories inscribed by complex local histories' (Benjamin 2008, p. 720).

But above all, the case of Zargaran shows how competitive dynamics acquire meaning and become intelligible only in the context of territorial transformation processes and their socioeconomic base. In the case of Bamiyan, the competitive dynamics can in fact be read as conflicting interests for land use amidst the valley's

---

[15] For the 'ethnic' orientation of the local government after 2001 see Adlparvar (2014).

[16] The change in the ethnic composition of the area is indicative in this regard. The ethnic composition in 1971 was: 60% Tajik, 30% Hazara, 10% Pashtun. At present, families belonging to the Hazara group (including Sadats) account for 88.6%, 11.2% belong to the Tajik group and 0.2% belong to other groups (mainly Pashtuns). 'The size of the Hazara group has increased greatly with new arrivals in the past few years: while over half of the Tajik population consists of families who settled in the area before 2000, most of the Hazara families (45.6%) arrived after 2010. Indeed, 94% of the families settling here after 2010 belong to the Hazara group' (LaGes 2018, pp. 74–75).

[17] Numerous accounts of the growing land-based interethnic conflict, with specific references to the case of Zargaran, can be found in Adlparvar (2014. For a reinterpretation of such dynamics, in terms of 'everyday ethnicity', see Adlparvar and Tadros (2016). For ethnic criteria in land distribution processes in 2004, see Alden Wiley (2004).

[18] Since 2001 the Afghan state has through various steps appropriated all uncultivated, dry land (*dasht*) above the highest irrigation channel in order to dispose of these vast areas for (political) redistribution, despite that land was used by the local communities as pasture land or for unirrigated agriculture (*lalmi*). For a detailed reconstruction until 2012 see Alden Wiley (2013). For the broader link between land appropriation and conflict see Adelkhah (2013).

transition from a traditional agro-pastoral system to a spatial system with an ever-growing number of typical urban functions, resulting in a pressing demand for space for residential and commercial destinations, as well as for transport infrastructure.

3.2 Tenure Security

Having a secure place to live considerably adds to one's quality of life and development as a human being. On one hand, the residents do not need to actively protect their claim over the land/house, which on the other hand, allows them to spend their time on other productive pursuits. However, what secure/tenure-security means is a complex affair.[19] In this regard, van Gelder (Gelder 2010) provides a useful tripartite view on tenure, which would account for:

3. *Perceived*: Tenure security as perceived by dwellers
4. *De jure*: Tenure security as a legal construct
5. *De facto*: Tenure security as (f)actually existing on ground.

These three aspects of tenure security are interdependent and there are several studies which push us to understand these differences as social constructs rather than a legal conflict (Ghertner 2008; Day 2008).

However, in the context of the modernization/bureaucratization of state organizations, guidelines linking land tenure to legal titles have also been established in Afghanistan, both on the side of the authorities, who enact it through the process of land titling, and on the side of the inhabitants of the new settlements, who manifest it through adhering to this policy.

In all the Afghan cities where it has been implemented, the land titling policy has been preceded by a systematic UN-Habitat survey aimed at mapping land use, surveying plots and buildings and checking occupancy conditions, as a precondition for the *Ministry of Urban Development and Land* (MUDL) to issue an Occupancy Certificate (OC). This information is used to judge whether or not to grant the certificate based on a long list of conditions.[20]

Among the criteria adopted for the granting of OC, the most relevant to this study are the lack of ownership conflicts with regard to plots on governmental land, and the possession of some form of documentation proving ownership for plots on private

---

[19] UN-Habitat categorizes tenure systems as follows: freehold, delayed freehold, registered leasehold ownership, public rental, private rental, shared equity, co-operative tenure, customary ownership, religious tenure, intermediate tenure and non-formal (UN-Habitat 2008). There is, therefore, by now a consensus in the scientific literature to dissolve the formal–informal dichotomy into a continuum (Davis 2006, pp. 178f; Jones 2016, pp. 166 and 179).

[20] The guidelines for the law's implementation distinguished between plots on government land and plots on private land. In the first case, the crucial criteria for eligibility were: 1. plots less than 300 square metere; 2. building at least 15 years old (de facto reduced to 5 years); 3, no conflicts over ownership; 4. not inside or overlapping with historical zone; 5. not located in a planned area. In the second case the crucial criteria were: 1. plot less than 500 square metere; 2. documents proving ownership; 3. not inside or overlapping with historical zone.

land. Both the criteria and the manner of application, highlight the government's intention to practice a policy of legalizing titles and forms of tenure security in accordance with the previous traditional arrangement of the local community. The documentation we accessed in Bamiyan confirms the authorities' effort to include, as far as is possible in the new legal system, documents of land ownership, which would otherwise be difficult to ascribe to that sphere. As seen in the example in Fig. 3,[21] one of the most comprehensive among those accessed, information such as the date and the seller's ownership title to the property is included.

At the same time, the data we collected in Zargaran undoubtedly shows that it is the inhabitants of the new settlements themselves who are seeking legalization of their ownership title, both as a guarantee of a greater independence and control of the traditional local sociocultural and political context, and as a consolidation of their own status.

According to the data from the UN-Habitat systematic survey, up to the fall of the Islamic Republic of Afghanistan, in Zargaran an OC had been issued for 314 out of 1,347 plots, namely 23% of the total (90 plots have not been registered), against a percentage of 12% for the whole of Bamiyan.

No matter how complex tenure security might be in terms of perceived, factual or legal construct, the possibility—theoretical as it may be—of accessing a formal/legal land ownership title, introduced by the 2017 law, led to a 100% rise in the average real estate value in Zargaran (Amiri 2022). Naturally, the values differ depending on accessibility. Real estate values are particularly high for less steep, car-friendly areas in the neighbourhood, average for less steep areas inaccessible by car and lower for very steep areas.

In this way, the land titling policy led to the emergence of a real-estate market in Bamiyan/Zargaran, which has largely dealt with the more recent demand created by settlers from other provinces in Afghanistan and by students attending Bamiyan University.

The comparison of the data on families residing in Zargaran collected directly by LaGeS in 2017 and 2021 highlights a significant increase (from 15 to 34%) in families living in rented properties. The 2021 survey also shows that over half of these families (56%) have settled since 2019.

This confirms assumptions which support the assertion that 'with ongoing commodification, informal property markets seem to obey the same laws and principles of any other market' (Boanada-Fuchs and Boanada-Fuchs 2018, p. 239).

---

[21] Translation: 'The reason for writing this document is that I, NAME, son of NAME, original from Katoway village, Central Bamiyan have sold a residential plot in Koshkak Valley to NAME, son of NAME, original from Yakawlang District, who now lives in Central Bamiyan. The price of this plot is 150.000 (one hundred fifty thousand) Afghanis. Half of the amount is 75.000 Afghanis. I received the entire sum in cash. The plot is surrounded as follows: to the east lies the land of NAME, to the west the land of NAME, to the north lies street 6 m, to the south a public street. I sell this plot to NAME, with witnesses. If anyone holds a claim upon this possession, I will be accountable. The dimension of the plot is 20 × 16 m'. SIGNATURE. As evident, this document somehow imitates legal language, but a date of the transaction, for example, is missing. The price of the transaction amounts to ca. 1900 $, current exchange rate.

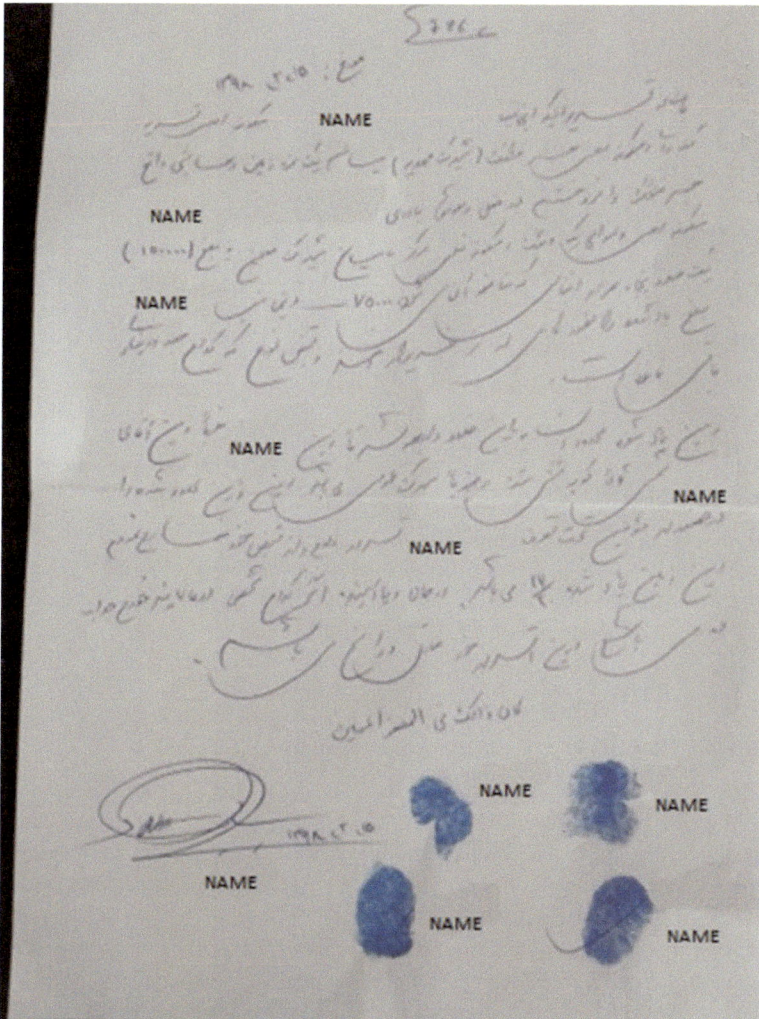

Fig. 3 Example of document proving purchase and sale of land

Therefore, there is a real danger that increased tenure security for owners could result in less security for tenants, particularly in a context of economic fragility such as the one in question. As can be seen by the distribution of the Zargaran population by employment status, the percentage of those not in work (school pupils, housewives, students, etc.) is very high and the economic burden of the family falls on a small number of family members, while unemployment and casual work are considerably high (Fig. 4).

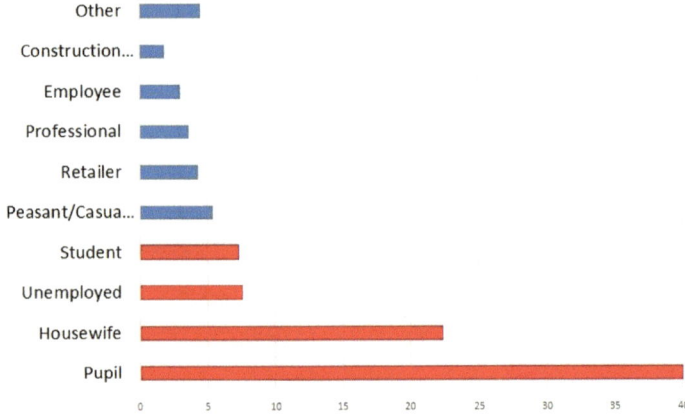

Fig. 4 Employment condition of Zargaran's population (*Source* authors' processing of LaGeS Household survey data 2021; N = 2,660)

This tells us that tenure security policies cannot automatically generate economic stability for the population. Indeed, they should be accompanied by measures which support the weaker segments of the population.[22]

4 Improvement of Housing and Urban Quality

As for the potential of the titling policy to improve housing and urban conditions, the first consideration is that, in psychological terms, tenure security is considered an important safety factor by the population. As such, this policy boasts widespread approval among the inhabitants (Fig. 5).[23]

Moreover, in terms of the connection between titling policies and urban development, an increase in tenure security strengthens the possibility of using the occupancy certificate question to launch urban regeneration strategies. This is first of all true at the individual level: access to an OC is correlated to higher rates of upgrading work on single properties (Table 1).

At least to some extent, the tendency of families who currently do not have the certificate to carry out upgrading work in the future can be interpreted as an expectation that one might be obtained later on. We do not have data on whether this has triggered a credit market (as hoped for by De Soto, for example De Soto (2001), but it is likely that the upgrading work was carried out personally, with the main expense being the building materials.

---

[22] For the danger of gentrification through the market see Davis (2006) and Amiri (2022), Leaf (1992), Kumar (1996), Mahadevia and Gogoi (2011).

[23] This has already been observed on a worldwide scale by De Soto (2001, p. 188); see also Varley (2017).

Fig. 5 Percentage of households' agreement with titling policy (*Source* authors' processing of LaGeS Household survey data 2021, N = 388)

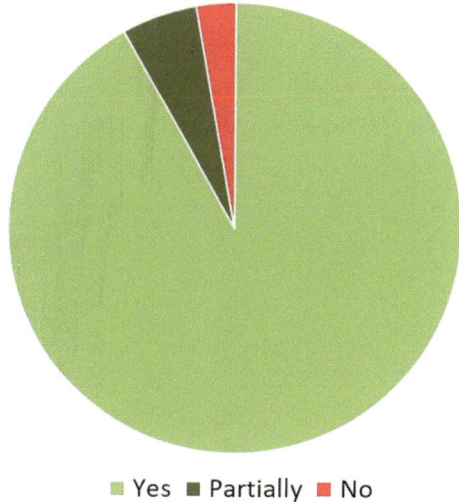

■ Yes ■ Partially ■ No

Table 1 Percentage of households by house improvement

Occupancy certificate	Yes, I have %	Yes, I will %	No %	N
Yes	39.1	47.7	13.2	243
No	6.5	73.8	19.6	107
Total	29.1	55.7	15.1	350

(*Source* authors' processing of LaGeS Household survey data 2021)

What is even more significant is that there is a positive correlation between OCs and people's willingness to contribute to neighbourhood regeneration projects, even by offering up portions of their own land (Table 2).

This fact provides a blueprint for urban upgrading policies, especially considering that, despite the rapid and extensive growth of new settlements, in the ethnically homogeneous neighbourhood there is a high degree of social capital: there are no conflicts, half the families declare that they have relations with other residents, while

Table 2 Percentage of households by willingness to contribute with land for benefit of neighbourhood

Ready to contribute with land for benefit of neighbourhood			
Occupancy certificate	Yes %	No %	N
Yes	47.1	52.9	240
No	28.0	72.0	107
Total	41.2	58.8	347

(*Source* authors' processing of LaGeS Household survey data 2021)

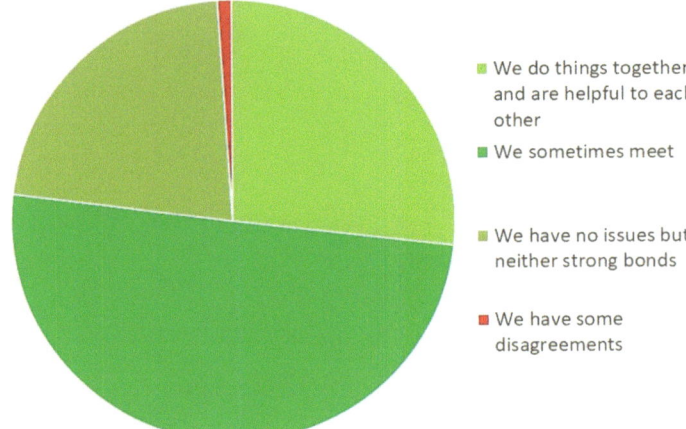

We do things together and are helpful to each other

We sometimes meet

We have no issues but neither strong bonds

We have some disagreements

Fig. 6 Quality of social relations in Zargaran (*Source* authors' processing of LaGeS Household survey data 2021, N = 388)

one quarter of families can also count on considerable social support in the local area (Fig. 6).

According to Harris' categories, Zargaran could thus be classified as an 'embedded' informal settlement defined by popular legitimacy, physical concentration and internal cooperation.[24]

Therefore, the social context is favourable towards the development of participatory urban regeneration projects, which could mobilize the regional resources of around half of the families with an OC; furthermore, 58% of families declare that they would participate with their own financial resources (a less binding solution than making their land available, but interesting all the same).[25] At least up to now, the land titling campaign in Zargaran does not confirm concerns about the inhabitants' tendency to isolation and increasing selfishness observed in other contexts.[26]

On the contrary, we have recently observed interesting examples of cooperation among the households, beyond assistance with material improvements. This can be seen in the financial cooperation of the households in building a new school, so as to improve the accessibility of educational facilities for pupils living in the northern part of Zargaran (Fig. 7).

---

[24] Harris (2018, p. 267) has distinguished five "modes of informality": latent, diffuse, embedded, overt, and dominant.

[25] Informal settlements in Afghanistan are comparatively little studied. Sahab and Kaneda, however, found out that the informal quarters of Kabul have fewer participatory mechanisms than formal ones (Sahab and Kaneda 2015).

[26] For the Latin American context see Aramburo Guevara (2018, pp. 85) and 96), Mion (2018, p. 102).

Fig. 7 One of the new schools in Zargaran run by the community (*Photo* M. Loda 2023)

5 Titling Policies in the Presence of Prestigious Cultural Heritage

As for the implementation of titling policies in the presence of prestigious cultural heritage, there is no doubt that Bamiyan is a paradigmatic case. The presence of exceptionally valuable cultural heritage in the area, and the valley's inclusion on the UNESCO list of World Heritage in danger, poses a particular problem concerning the implementation of land titling policies.

As can be seen from the map in Fig. 8, until the fall of the Afghan Republic, OCs in Zargaran were only issued to the north of an imaginary line dividing the neighbourhood from east to west. Below this line, the plots fall partly into the so-called 'planned area'.[27] These are the plots of land provided by the municipality between 2003 and 2010 in the south-eastern corner of Zargaran, for which the owners had already paid sums of money on various counts. In order to regulate these plots, an agreement is being drawn up between the MUDL, central authorities and the municipality.

Below the e on the protection of historic and east–west line where OCs are not issued, there are also plots that did not receive the OCs because they fall into the

---

[27] No new occupancy certificates were issued by the new government. However, those issued by the previous government remained in force.

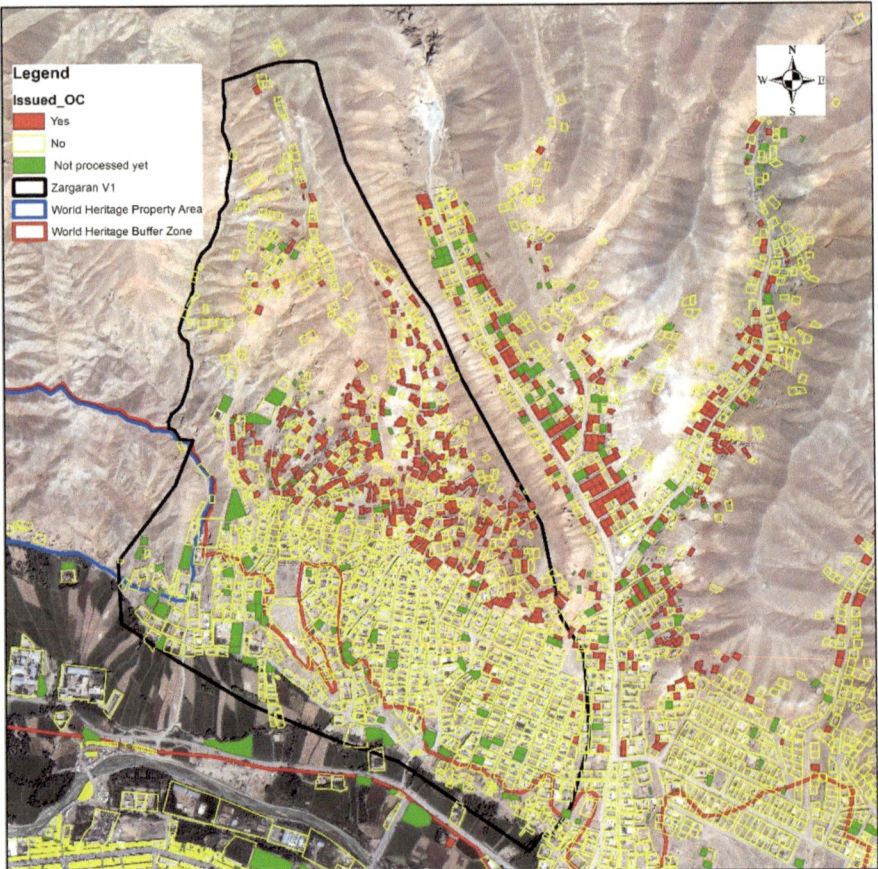

Fig. 8 Issued OC (*Source* authors' processing of UN-Habitat data, unpublished)

historical zone established by the 'Afghan Law on the Protection of Historic and Cultural Properties' (2004). The historical (or archaeological) zone includes the buffer zone that was drawn around the World Heritage property of the Buddha niches the year before, as well as a wide area north of it. In this extended area, the lack of OCs is justified by respect for the 'Not inside or overlapping with historical zone' criterion (see note 19).

By this way, the very vast and densely populated area corresponding to the buffer zone was therefore excluded from land titling policy. The very vast and densely populated area corresponding to the buffer zone was therefore excluded from land titling policy. However, a buffer zone aims at safeguarding the view/visibility of a property and at protecting the heritage from potentially unchecked initiatives. Considering that the regulations placed by the *Bamiyan Strategic Master Plan* approved in 2018 (LaGes 2018)—particularly the severe building work restrictions on this buffer zone—serve exactly this goal, it needs to be evaluated whether excluding such a large built-up area from land titling policy leads to the failure of a large number of tendentially privileged owners (residents in gently sloping areas that can be reached by car) to participate in urban upgrading projects.

6 Final Observations

The data examined in the case of Bamiyan/Zargaran leads to the conclusion that, on the whole, land titling policies definitely meet the approval of the population and provide a tool that can facilitate participatory urban upgrading processes.

This is true both at the individual level of building upgrading work by single households and above all at the collective level, as it encourages an openness to active participation—by making portions of land and/or financial resources available—in neighbourhood upgrading projects. As for the possibility of implementing urban regeneration projects through land titling policies, the quality of the social capital is definitely an encouraging factor: the lack of specific conflicts and the networks of interactive relationships and mutual support among families help to boost a sense of trust in collective initiatives.

Land titling policies have also shown that they act on land values by contributing to the development of a real estate market. However positive this may be in the future in terms of local economic enterprise, this mechanism could have an impact on economically fragile contexts such as Zargaran, making housing inaccessible and effectively expelling the poorest segments of the population (eviction). Therefore, this calls for the land titling policy to be accompanied by measures to support the more fragile sectors of the population.

Lastly, the case of Bamiyan/Zargaran leads to some reflections on the implementation of land titling policies in contexts of prestigious cultural heritage. Despite agreeing with the precautionary concern resulting in the failure to issue OCs to plots of land included in or alongside historical areas, the exclusion of all plots in the Zargaran buffer zone from OCs undermines a large reserve of resources that could be used to upgrade buildings and urban areas. In the case of such extensive and densely populated areas as the Zargaran buffer zone, it must be assessed whether the ban on issuing OCs can be replaced by specific agreements to issue OCs to single individuals and the local community.

Credit authorship contribution statement. Mirella Loda: Conceptualization, Methodology, Supervision, Formal Analysis, Writing—Original Draft, Visualization, Writing, Review & Editing. Nipesh Narat Narayanan: Conceptualization. Bashir Amiri: Data curation, Formal Analysis.

References

Abdulai A, Owusu V, Goetz R (2011) Land tenure differences and investment in land improvement measures: theoretical and empirical analyses. J Dev Econ 96(1):66–78

Adelkhah F (2013) Terre et guerre en Afghanistan. Presses Univ. de Provence, Aix-en-Provence

Adlparvar N (2014) When glass breaks, it becomes sharper: de-constructing ethnicity in the Bamiyan-Valley, Afghanistan. PhD thesis, Institute of Development Studies, University of Sussex

Adlparvar N, Tadros M (2016) The evolution of ethnicity theory: intersectionality, geopolitics and development. Inst Dev Stud, IDS Bull 47(2):123–136

Ahmadi AR (2018) Rehabilitation of informal settlements in Bamiyan–the neighbourhood of Zargaran. MA thesis, Master in Urban Analysis and Management, Florence University

Alden Wiley L (2004) Land relations in Bamiyan province. Findings from a 15 village case study. Afghanistan Research and Evaluation Unit (AREU), Kabul

Alden Wiley L (2013) Land, people and the state in Afghanistan: 2002–2012, Afghanistan. Research and Evaluation Unit (AREU), Kabul

Ali DA, Deininger K, Goldstein M (2014) Environmental and gender impacts of land tenure regularization in Africa: pilot evidence from Rwanda. J Dev Econ 110:262–275. (Land and property rights)

AlSayyad N (2004) Urban informality as a "New" way of life. In: AlSayyad N, Roy A (eds) Urban infromality. Transnational perspectives from the Middle East, Latin America, and South Asia, pp 7–29. Lexington Books, Lanham

Amiri B (2022) Land titling: an opportunity for socio-economic development of informal settlements? The case of Zargaran. MA thesis, Florence University

Amiri B, Lukumwena A (2018) An overview of informal settlement upgrading strategies in Kabul city and the need for an integrated multi-sector upgrading model. Curr Urban Stud 348–365

Anwar NH (2012) State power, civic participation and the urban frontier: the politics of the commons in Karachi. Antipode 443:601–620

Aramburo Guevara NK (2018) the role of the state in slum improvement: a critical examination of a Peripheral Barrio of Lima. In: Petrillo A, Bellaviti P (eds) Sustainable urban development and globalization. New strategies for new challenges-with a focus on the global south, pp 85–98. Springer, Cham

Arnott R (2008) Housing policy in developing countries. The importance of the informal economy. World Bank, Washington D.C

Balakrishnan S, Pani N (2021) Real estate politicians in India. Urban Stud 58(10):2079–2094

Bangasser PE (2000) The ILO and the informal sector. An institutional history. International Labour Organization

Bayat A (2000) From 'Dangerous Classes' to 'Quiet Rebels'. Politics of the urban subaltern in the global south. Int Sociol 15(3):533–537

Benanav A (2019) The originis of informality. The ILO at the limit of the concept of unemployment. J Glob Hist 14(1):107–125

Benjamin S (2008) Occupancy urbanism. Radicalizing politics and economy beyond policy and programs: debates and developments. Int J Urban Reg Res 32(3):719–729

Boanada-Fuchs A, Boanada-Fuchs V (2018) Towards a taxonomic understanding of informality. Int Dev Plan Rev 40(4):397–420

Bredenoord J, van Lindert P (2010) Pro-poor housing policies. rethinking the potential of assisted self-help housing. Habitat Int 34:278–287

Bromley R (1990) A new path to development? The significance and impact of Hernando de Soto's ideas on underdevelopment, production, and reproduction. Econ Geogr 66(4):328–348

Calogero P, Schütte S (2018) Informalität von oben und unten. Stadtenwicklung in Kabul im Kontext von Staatsaufbau und militärisch-humanitärer Intervention, sub/urban. Zeitschrift für kritische Stadtforschung 6(2/3):7–30

Cheng W, Xu Y, Zhou N et al (2019) How did land titling affect China's rural land rental market? Size, composition and efficiency. Land Use Policy 82:609–619

Clark K (2023) From land-grabbing to haircuts: The decrees and edicts of the Taleban supreme leader. Afghan Analysts Network, https://www.afghanistan-analysts.org/en/reports/rights-fre edom/from-land-grabbing-to-haircuts-the-decrees-and-edicts-of-the-taleban-supreme-leader/. Last Accessed 16 July 2023

Das AK, Takahashi LM (2009) Evolving institutional arrangements, scaling up, and sustainability. issues in participatory slum upgrading in Ahmedabad, India. J Plan Educ Res 29:213–232

Davis M (2006) Planet of slums. Verso, New York

Day J (2020) Sister communities: rejecting labels of informality and peripherality in Vanuatu. Int J Urban Reg Res 44(6):989–1005

De Soto H (2001) The mystery of capital: why capitalism triumphs in the west and fails everywhere else. Black Swan Books, London

De Janvry A, Emerick K, Gonzalez-Navarro M et al (2015) Delinking land rights from land use: certification and migration in Mexico. Am Econ Rev 105(10):3125–3149

Deininger K, Jin S, Xia F et al (2014) Moving off the farm: land institutions to facilitate structural transformation and agricultural productivity growth in China. World Dev 59:505–520

Del Mistro R, Hensher DA (2012) Upgrading informal settlements in South Africa: policy, rhetoric and what residents really value. Hous Stud 24(3):333–354

Devkar G, Thillai Rajan A, Narsyanan S (2019) Provision of basic services in slums: a review of the evidence on top-down and bottom-up approaches. Dev Policy Rev 37:3331–3347

Dovey K (2012) Informal settlement and complex adaptive assemblage. Int Dev Plan Rev 34(3):371–390

Dowell DE, Clarke G (1996) A framework for reforming urban land policies in developing countries. International Bank for Reconstruction, Washington D.C

Durand-Lasserve A, Royston L (eds) (2002) Holding their ground. Secure land tenure for the urban poor in developing countries. Routledge, London

Fernándes E (2011) Regularization of informal settlements in Latin America. Lincoln Institute of Land Policy, Cambridge MA

Field E (2007) Entitled to work: urban property rights and labor supply in Peru. Q J Econ 122(4):1561–1602

Galiani S, Schargrodsky E (2010) Property rights for the poor: effects of land titling. J Public Econ 94(9):700–729

Ghertner DA (2015) Rule by aesthetics: world-class city making in Delhi. Oxford UP, New York

Ghertner DA (2020) Lively lands: the spatial reproduction squeeze and the failure of the urban imaginary. Int J Urban Reg Res 1468–2427:12926

Gilbert A (2002) On the mystery of capital and the myths of Hernando de Soto. Int Dev Plan Rev 24(1):1–19

Gilbert A (2007) The return of the slum: does language matter? Int J Urban Reg Res 31(4):697–713

Goldstein M, Udry C (2008) The profits of power: land rights and agricultural investment in Ghana. J Polit Econ 116(6):981–1022

Habib J (2011) Urban cohesiveness in Kabul city: challenges and threats. Int J Environ Stud 68(3):363–371

Harris R (2018) Modes of informal urban development: a global phenomenon. J Plan Lit 33(3):267–286

Heikkila EJ, Harten JG (2019) Can land use regulation be smarter? Planner's role in the informal housing challenge. J Plan Educ Res 1–12

Ho P, Spoor M (2006) Whose land? The political economy of land titling in transitional economies. Land Use Policy 23(4):580–587

Islamic Republic of Afghanistan (2019) Ministry of urban development and land, Afghanistan *Land Administration System Project* (ALASP). Social management framework, final report, https://documents1.worldbank.org/curated/zh/922661547187440452/SFG 4977-V2-REVISED-EA-P164762-PUBLIC-Disclosed-2-6-2019.pdf. Last Accessed 08 Dec 2023

Jones P (2016) Informal urbanism as a product of socio-cultural expression: insights from the Island Pacific. In: Attia S, Shabka S, Shafik Z, Ibrahim S (eds) Dynamics and resilience of informal areas. International perspectives, pp 165–186. Springer, Cham

Jones P (2017) Formalizing the informal: understanding the position of informal settlements and slums in sustainable urbanization policies and strategies in Bandung, Indonesia. Sustainability 1–27

Jonnalagadda I, Stock R, Misquitta K (2021) Titling as a contested process: conditional land rights and subaltern citizenship in South India. Int J Urban Reg Res 1468–2427:13002

Kellett P, Napier M (1995) Squatter architecture? A critical examination of vernacular theory and spontaneous settlement with reference to South America and South Africa. Tour Sustain Dev Rev 11:7–24

Koster M, Nuijten M (2016) Coproducing urban space: rethinking the formal/informal dichotomy. Singap J Trop Geogr 37:282–294

Kucina I (2018) Architectures of informality. Kia Series, Dessau

Kumar S (1996) Landlordism in third world urban low-income settlements: a case for further research. Urban Stud 33(4/5):753–782

La Mantia CI (2018) Humanizing urbanism. On embracing informality and the future of Johannesburg. In: Petrillo A, Bellaviti P (eds) Sustainable urban development and globalization. New strategies for new challenges-with a focus on the global south, pp 48–63. Springer, Cham

LaGes (Laboratorio di Geografia Sociale, Università degli Studi di Firenze) (2018) Bamiyan strategic master plan. Polistampa, Firenze

Leaf M (1992) Informality and urban land markets. Berkeley Plan J 7(1):132–138

Lizarralde G (2011) The invisible houses. Rethinking and designing low-cost housing in developing countries. Routledge, New York, London

Lutzoni L (2016) In-formalized urban space design. Rethinking the relationship between formal and informal. City, Territ Arch 30(2):1–14

Mahadevia D, Gogoi T (2011) Rental housing in informal settlements. A case-study of Rajkot, center for urban equity. Center for Enrironmental Planning and Technology, CEPT University, Working paper 14. Ahmedabad

Majale M (2008) Employment creation through participatory urban planning and slum upprading: the case of Kitale, Kenya. Habitat Int 32:270–282

McFarlane C (2020) Repenser l'informalité: La politique, les crises et la ville. Lien Social Et Politiques 76:44–76

Meinzen-Dick R, Mwangi E (2009) Cutting the web of interests: pitfalls of formalizing property rights. Land Use Policy 26(1):36–43. (Formalisation of land rights in the south)

Mion V (2018) Undergrowth urbanism: the role of user generated practices in the informal city. In: Petrillo A, Bellaviti P (eds) Sustainable urban development and globalization. New strategies for new challenges-with a focus on the global south, pp 99–116. Springer Cham

Mukhija V (2001) Upgrading housing settlements in developing countries. Cities 18(4):213–220

Opoko AP, Oluwatayo AA, Amole B et al (2020) How different actors shape the real estate market for informal settlements in Lagos. Environ Urban 371–388

Payne G, Durand-Lasserve A, Rakodi C (2009) The limits of land titling and home ownership. Environ Urban 21(2):443–462

Porter L (2011) Informality, the commons and the paradoxes of planning: concepts and debates for informality and planning. Plan Theory Pract 12(1):115–153

Riley E, Wakely P (2011) The case for incremental housing, cities alliance policy research and working paper series, no 1. Cities without Slums, Washington D.C

Roy A (2005) Urban informality. Toward an epistemology of planning. J Am Plan Assoc 71(2):147–158

Roy A (2011) Slumdog cities: rethinking subaltern urbanism. Int J Urban Reg Res 35(2):223–238. Here quoted as: Roy A (2012) Slumdog cities: rethinking subaltern urbanism. In: Angélil M, Hehl R (eds) Informalize! Essays on the political economy of urban form, vol 1, pp 107-142. ETH Zurich, Berlin

Russ LW, Takahashi LM (2013) Exploring the influence of participation on programme satisfaction: lessons from the Ahmedabad slum networking project. Urban Stud 50(4):691–703

Sahab S, Kaneda T (2015) A study on neighborhood functions of 'Gozars' in Kabul, Afghanistan. J Arch Plan 80(216):2253–2260

Schindler S (2017) Towards a paradigm of southern urbanism. City 21(1):47–64

Simone A (2001) On the worlding of African cites. Afr Stud Rev 44(2):15–41

Toulmin C (2009) Securing land and property rights in sub-Saharan Africa: the role of local institutions. Land Use Policy 26(1):10–19. (Formalisation of land rights in the south)

Turner JFC (1967) Barriers and channels for housing development in modernizing countries. J Am Inst Plann 33:167–180

Turner JFC (1983) From central provision to local enablement. New directions for housing policies. Habitat Int 7(5/6):207–210

UN-Habitat (2008) Secure land-rights for all. Nairobi

UN-Habitat (2015) United Nations settlement programme, habitat III issue paper 22. Informal Settlements, Nairobi

UN-Habitat (2015) State of Afghan cities (SoAC), vol 2. Nairobi, Kabul

UN-Habitat (2017) Afghanistan housing profile. Nairobi

Upadhya C, Rao DM (2022) Dispossession without displacement: producing property through slum redevelopment in Bengaluru, India. Environ Plan A: Econ Space 1–17

Van Gelder J-L (2010) What tenure security? The case for a tripartite view. Land Use Policy 27(2):449–456

Varley A (2017) Property titles and the urban poor: from informality to displacement? Plan Theory Pract 18(3):385–404

Wani ZA (2021) Afghanistan's Neo-Taliban puzzle. South Asia Res 41(2):220–237

Ward C (2021) Land financialisation, planning informalisation and gentrification as statecraft in Antwerp. Urban Stud 1–18

Watson V (2009) Seeing from the south: refocussing urban planning on the globe's central issues. Urban Stud 46(1):2259–2275

Rehabilitation of Informal Settlements in Heritage Sites: Zargaran (Bamiyan)*

Mirella Loda⊙, Gaetano Di Benedetto, and Giovanna Potestà⊙

Abstract This paper outlines the main points of the rehabilitation plan for the informal settlement of Zargaran, in the Bamiyan Valley, which the University of Florence has been developing with funding from the Italian Development Cooperation. The plan has been prepared through a careful analysis of the area and building conditions. The plan aims to provide general guidance on the rehabilitation of informal housing settlements across UNESCO sites, not least because of the overlap between the westernmost part of the village and the UNESCO property of the Buddha Cliff.

Keywords Informal settlements · Heritage sites · Bamiyan

1 Introduction

In the context of the vast process of urbanization of the population occurring across the Global South and which, according to UN-Habitat data (UN-Habitat 2015a), will bring the urbanization rate of the Afghan population from 24% in 2017 to 50% in 2060, Bamiyan is where the phenomenon is manifested most strongly throughout Afghanistan.

Bolstered by the fall of the first Taliban regime, the valley—due to the addition of relocations from the countryside and the return of refugees (in a context of a very

---

* The authors have contributed to the present essay as follows: Mirella Loda, paragraphs 1, 2, 3; Gaetano Di Benedetto and Giovanna Potestà, paragraph 4.

M. Loda
Department of History, Archaeology, Geography, Fine and Performing Arts (SAGAS) and Laboratory for Social Geography (LaGeS), University of Florence, Via San Gallo 10, 50129 Firenze, Italy
e-mail: mirella.loda@unifi.it

G. Di Benedetto · G. Potestà (✉)
Laboratory for Social Geography (LaGeS), University of Florence, Via San Gallo 10, 50129 Firenze, Italy
e-mail: giovanna.potesta@unifi.it

153

high birth rate of 36/1,000 inhabitants (Indexmundi 2023))—saw its population rise from 7,300 in 1979 to 52,000 in 2017 (LaGeS 2018).

In Bamiyan, as in the rest of Afghanistan, population growth has naturally been accompanied by a marked increase in housing demand, the expansion of spontaneous housing, and the proliferation of various forms of informal settlements.

According to the concise definition provided by UN-Habitat (2015) informal settlements are "residential areas where:

1. inhabitants have no security of tenure vis-à-vis the land or dwellings they inhabit, with modalities ranging from squatting to informal rental housing
2. the neighborhoods usually lack, or are cut off from, basic services and city infrastructure
3. the housing may not comply with current planning and building regulations and is often situated in geographically and environmentally hazardous areas" (UN-Habitat 2015b).

In the Bamiyan Valley numerous settlements exhibit (one or more of) these characteristics.

The village of Zargaran is just one of the informal settlements identified in Bamiyan. Zargaran is located on the northern side of the central Bamiyan valley, directly to the east of the most important UNESCO site, the great "Buddha cliff", and to the west of Jugra Khail (Fig. 1).

As explained in chapter "Improving Urban Quality Through Land Titling? Considerations from the Bamiyan Case" of this volume, until 2003 Zargaran had consisted

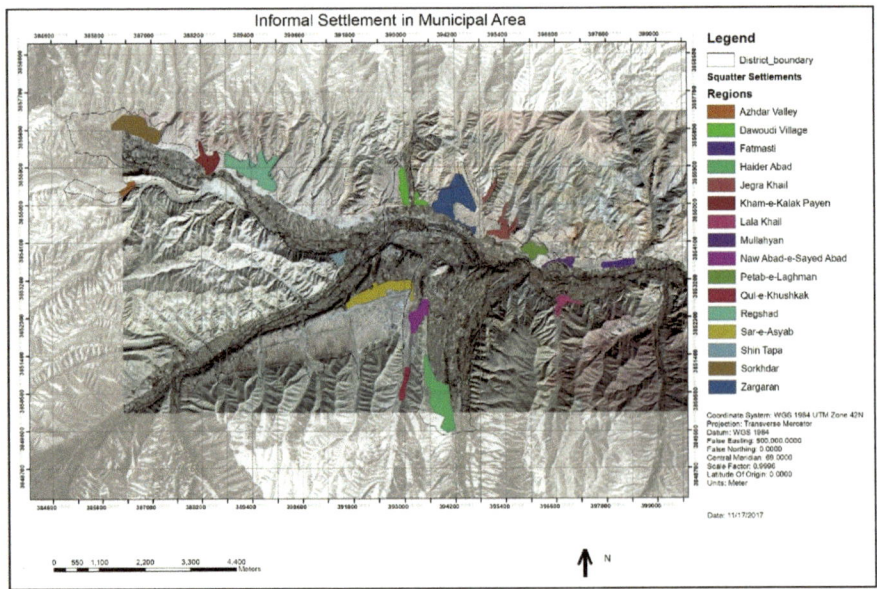

Fig. 1 Informal settlements in the Bamiyan Valley (*Source* Ahmadi 2018)

of only a few farmhouses built on the hill above the highest irrigation channel. The area was used as a commons by the adjacent village of Dawoodi, to the west. In 2002 the new municipality started assigning 224 plots of 300–400 square meters to incoming refugees, each plot in the south-eastern corner of the present-day Zargaran. The small, planned area is easily recognizable as a result of its regular street grid. This came to an end in 2007, but over the same years many people started spontaneously settling in the neighbourhood (Ahmadi 2018), with a notable increase in recent years. The spontaneous settlement that developed around the original core now presents many problems with regard to infrastructure and service provision.

The challenges observed in Zargaran are typical of many informal settlements in the Bamiyan area and, indeed, throughout the country. Hence the decision by Unifi and AICS to develop a model plan for Zargaran, aimed at upgrading informal dwellings so that they meet appropriate building standards, as well as the overall rehabilitation of the settlement. Zargaran's rehabilitation plan could thus become a paradigm for the management of other informal settlements in Afghanistan and similar realities.

Moreover, given the overlap between the extreme western part of the village and the UNESCO property comprising the Buddha cliff, Zargaran can funtion as an exemplary case in helping to define objectives and strategies for upgrading (informal) housing settlements in UNESCO sites.

Considering the UN-Habitat definition of informal settlements mentioned above, criteria 1 has been dealt with in chapter "Improving Urban Quality Through Land Titling? Considerations from the Bamiyan Case" of the present volume. This contribution concentrates on criteria 2 and 3, concerning interventions on the building and infrastructure aspects of informal settlements more closely.

2 Analytical Approach

To ensure that the plan's objectives adhere to the complexity of the problems on site, a systematic and up-to-date collection of information on the socio-territorial context was first carried out, acknowledging that the variables considered relevant to the definition of an upgrading plan are as follows:

- residential density (residents per hectare)
- land ownership
- household income
- age of buildings
- condition of buildings
- demolition or relocation
- water supply, sanitation, electricity and waste management
- environmental hazards
- public facilities (schools, healthcare, religious buildings, etc.)
- regulation and zoning for residential buildings

– attitude of residents towards improvement.

As for the urban architectural part, a rehabilitation plan needs to identify and define the following categories:

– size of roads in relation to the city and internal network
– building type and height of construction
– public buildings or areas where public buildings could be built
– buildings in need of demolition, reconstruction, rehabilitation or regularization according to the building codes of the Master Plan
– cadastral list of properties necessitating expropriation
– areas adjacent to public buildings necessary to their present and future utilization.

The collection of such information was carried out:

(a) Using data made available by UN-Habitat as a result of the survey carried out on the built-up area of the village (as well as in the rest of the valley) from 2019 to 2021 within the framework of the Afghan land-titling policies (see chapter "Framing Planning and Conservation Activities in the Local Socio-Cultural Context: Ethnicity and Gender in Bamiyan"). This data has been surveyed for other purposes, but contains precious information about plot size, house layout and materials etc., for (nearly[1]) all buildings in Zargaran.
(b) By direct survey. Specifically, by carrying out a household and building survey, conducted by LaGeS in 2017 across the entire Bamiyan Valley, drawing on a sample of 25 percent of households, and through a second household survey conducted by LaGeS in Zargaran in the spring of 2021, on 50 percent of resident households. In addition to providing an up-to-date picture of the socioeconomic characteristics of the population (briefly discussed in chapter "Framing Planning and Conservation Activities in the Local Socio-Cultural Context: Ethnicity and Gender in Bamiyan"), the surveys provided information on the technical-building characteristics of dwellings and housing patterns. Comparing LaGeS data from 2017 and 2021 made it possible to highlight and quantify important transformational processes taking place in the area (such as population growth, built-up area, and so on). In addition, the data obtained served to complement the information on housing characteristics made available by the UN-Habitat survey.
(c) At the Municipality and at the local MUDL office.

Based on the data acquired, we now proceed to illustrate the situation in Zargaran.

---

[1] See footnote 2.

3   The Local Context

3.1   *Socio-Demographic Context*

The village of Zargaran—according to a perimeter that follows to the west the valley furrow leading off from the central cliff containing the Buddha niches, and to the east the ridge leading off from the village of Jugra Khail—spans an area of 109.68 hectares. The area has recently undergone significant population growth and intense construction.

Zargaran's population, estimated at 6,604 inhabitants in May 2017 (first LaGeS household survey, LaGeS 2018), grew to 8,866 inh. in 2021 (second LaGeS household survey). In only four years there was a 34 percent increase in the population, with 22 percent of the households settling between 2019 and 2021.

Population density—which still has the highest values in the entire valley, with peaks of more than 110 inh./ha in the central part (LaGeS 2018, p. 74)—has increased on average from 60.58 inh./ha in 2017 to 81.31 inh./ha in 2021.

Eighty percent of the population settled in Zargaran are IDPs from rural areas of the central Bamiyan District or the rest of the province; 13% are from other provinces and 3% are refugees who returned to Afghanistan after the fall of the first Taliban regime.

Job opportunities related to the city's growth and bazaar are the main pull factor, cited first in determining settlement choice by just under half of the surveyed population; security reasons ranked first for just under ¼ of the sample (Fig. 2).

Despite expectations regarding job opportunities offered by the urban-oriented development of Bamiyan, the economic situation of the population of Zargaran is in fact rather precarious. 72% of the population is in non-professional status while 11% is unemployed or in temporary employment. In 12% of cases, part of the income comes from activities carried out within the household (mainly from handicrafts, 7%, and tailoring, 5%).

The average monthly income per household is $161 (the per capita income is $25). In a very weak economic environment, households with higher incomes tend

Fig. 2   Area of origin (left) and reason for settling in Zargaran (right) (*Source* author's processing of data from LaGeS survey 2021)

to be concentrated in the southeastern and flat part of the village, while those with lower incomes in the impervious northern areas of more recent urbanization.

In slightly more than ¼ of cases, household income is supplemented by raising animals for self-consumption: mainly poultry, but in some cases (6%) also cows, sheep and donkeys. 1/5 of households have a small vegetable garden.

Seventy one percent of occupied houses are owned, while 25% of the houses are rented.

3.2 The Quality of the Buildings

Alongside the population increase in the Zargaran area there has been significant urban expansion. There were 633 buildings captured by satellite in 2015 (UN-Habitat (Ahmadi 2018))—1,050 in June 2017 (LaGeS 2018). In June 2021, according to the building survey conducted for the implementation of the land titling campaign (Djamil 2016), UN-Habitat recorded 1,437 parcels. According to the data collected on 1,366 of them, 1,310 were residential buildings, 15 business buildings and 7 were mixed-purpose.[2] Apart from the school and the cemetery area, there were essentially no public outdoor spaces in the neighborhood (Fig. 3).

The settlement has continued to expand in line with a settlement model that traditionally preserves the fertile soil of the valley floor; the new buildings, however, have been built up steep mountain slopes, which poses serious environmental issues, as well as problems with regard to the provision of services (Fig. 4).

Most buildings (90%) are single-storey houses, 59 are two-storey houses and only 2 buildings consist of 3 floors (Djamil 2016). These are 94% raw mud-bricks buildings, 2% concrete buildings, and 3.7% mixed-material buildings, made of concrete and mud-bricks.

The raw mud bricks used in the construction of most houses, ideally above a pebbled groundwork, is the traditional building material in the area, used both for private dwellings and for public buildings. As a building material, mud bricks are advantageous for several reasons (Djamil 2016; USAID 2022; Sruthi 2013):

(1) Clay, the raw material, is available everywhere. It is cost-effective, since it requires no transportation and is cheap to produce.
(2) Mud bricks are, if properly maintained, highly durable.
(3) Mud bricks have excellent temperature insulation qualities both against the heat and against the cold, the latter particularly relevant to Bamiyan.
(4) Mud brick houses, being of the same material and color, blend into the surrounding landscape perfectly.

The problem with mud bricks lies in their permeability. They must, therefore, be properly maintained, i.e. coated with loam plaster mixed with straw (*pakhsa*), in

---

[2] A control analysis conducted using Basemap images on October 10, 2023 showed that an additional 34 plots were built in the period between the UN-Habitat 2021 survey and 2023.

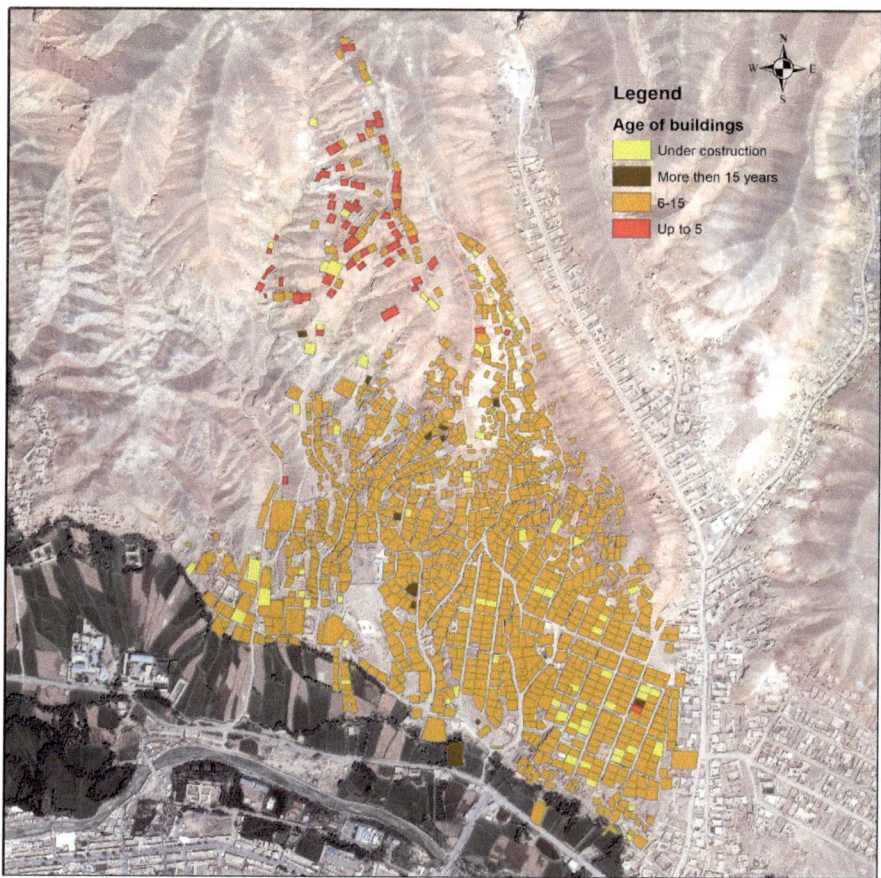

Fig. 3 Age of buildings (*Source* author's processing of UN-Habitat data 2021)

order to be protected from humidity. Since the loam plaster can get washed off by rainfall, it has to be renewed every two years on average. This is most likely the reason why better-off families in Zargaran and elsewhere opt for concrete as their preferred building material, alongside its "modernist appeal".

Although highly homogeneous in terms of material and construction techniques, the buildings differ significantly in volume, with a gradual, progressive reduction in volume in the more recently built dwellings in the northern part of the district (Fig. 5).

Fig. 4 Buildings on steep slopes (*Source* author's processing of UN-Habitat data 2021)

3.3 Equipment, Utilities and Facilities

Data collected in 2017 through the LaGeS Household survey provides information regarding the endowments available for buildings at that time. As we will see shortly, the situation regarding available endowments has improved somewhat in the last few years.

This is particularly the case for drinking water, which is made available in Zargaran by a network installed and operated by private companies on behalf of the Community Council (Shura) with funding from NGOs. In 2017, drinking water was pumped from two wells into three reservoirs within the neighborhood and then distributed to the households. Drinking water was available in 81% of the buildings[3] while only 3%

---

[3] In this respect, the situation in Zargaran was better than in the rest of the valley, where only 55 percent of households received drinking water. At present, however, a project for building a

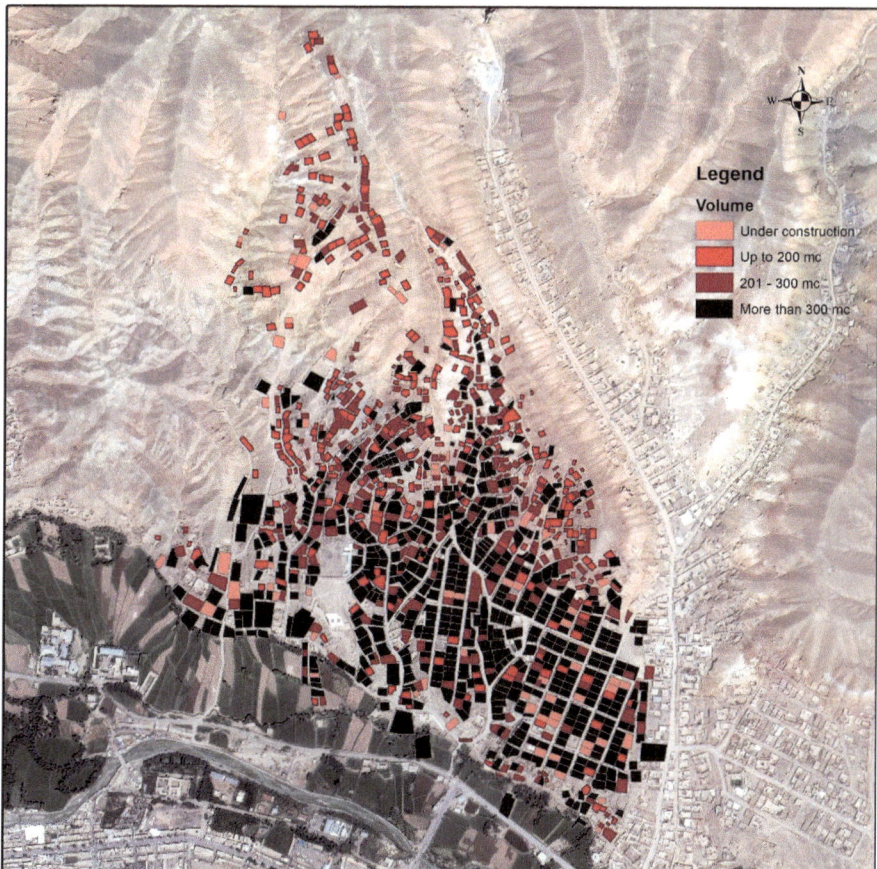

Fig. 5 Buildings by available volume (*Source* author's processing of UN-Habitat data 2021)

had their own well; a critical situation affected the new buildings in the western and northern parts of the village, especially those constructed on steep slopes. According to information directly gathered on site in the spring of 2023, all buildings in Zargaran have since been connected to the drinking water network, thanks to the installation of a fourth water reservoir built on an elevated point at the extreme northern tip of the settlement (Fig. 6).[4]

---

water pipe from Ahangaran Valley (3700 m) to Dasht-e Essa Khan is being prepared, which should substantially improve water supply, even in the southern part of the valley. The cost of the project is estimated at $1 million.

[4] The case of Zargaran draws attention to possible perverse effects generated by direct cooperative support to local communities when not framed within a more general settlement management strategy involving local administrations. In Zargaran, the laudable intent of the NGOs to expand access to the water supply in a direct relationship with local communities has in fact pandered to further spontaneous expansion of the built-up area in vacant but highly impervious parts of the area,

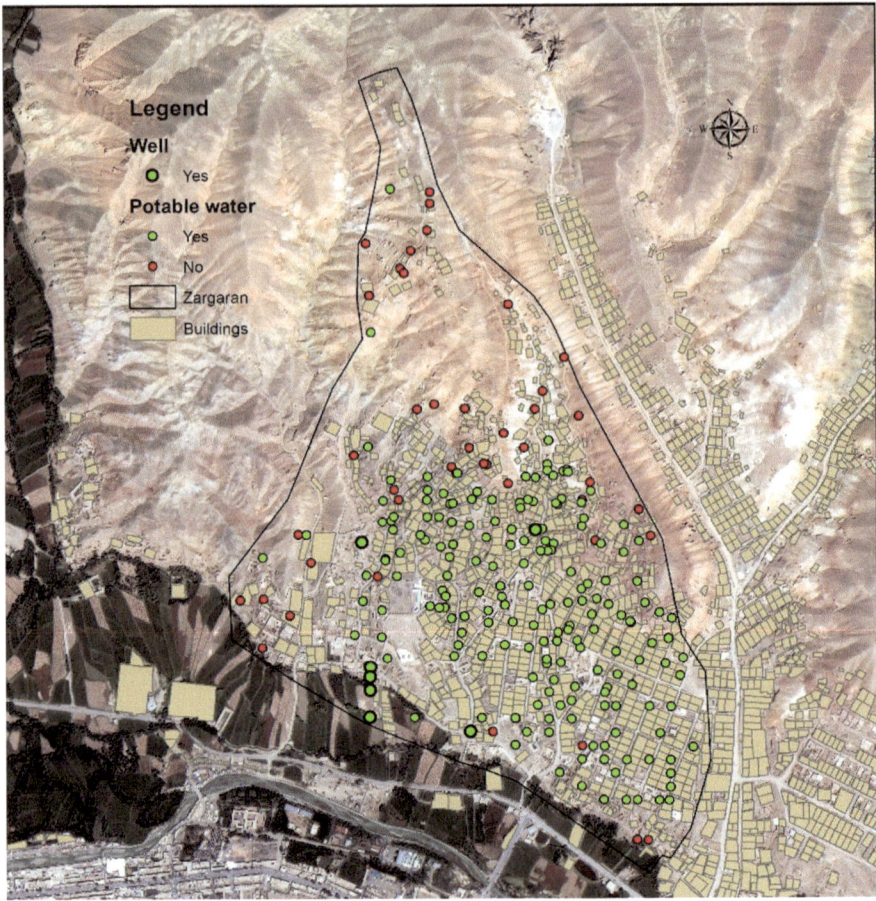

Fig. 6 Potable water and wells (*Source* author's processing of LaGeS Household survey data 2017)

Bamiyan and Zargaran are generally provided with electric power through the TUTAP[5] power line. 76 percent of the buildings are connected to the electricity grid. Most buildings not connected to the grid are concentrated in the western and northern parts, and are considered the under-equipped parts of the district (Fig. 7).[6]

Zargaran's sewage situation is critical. Only 1% of the buildings have a septic tank and about 52% have a soakage well. About 46% of the buildings do not have a proper sewage cleaning system, most prominently in the western and northern parts

---

where the local government is unlikely to be able to integrate other basic services (sewage cleaning and road networks).

[5] TUTAP is a power transmission line set to link the energy-rich states of Turkmenistan, Uzbekistan and Tajikistan, with Afghanistan and Pakistan. The project, started in 2013, was backed by the Asia Development Bank (ADB).

[6] There is no known data on the situation having improved, compared to 2017.

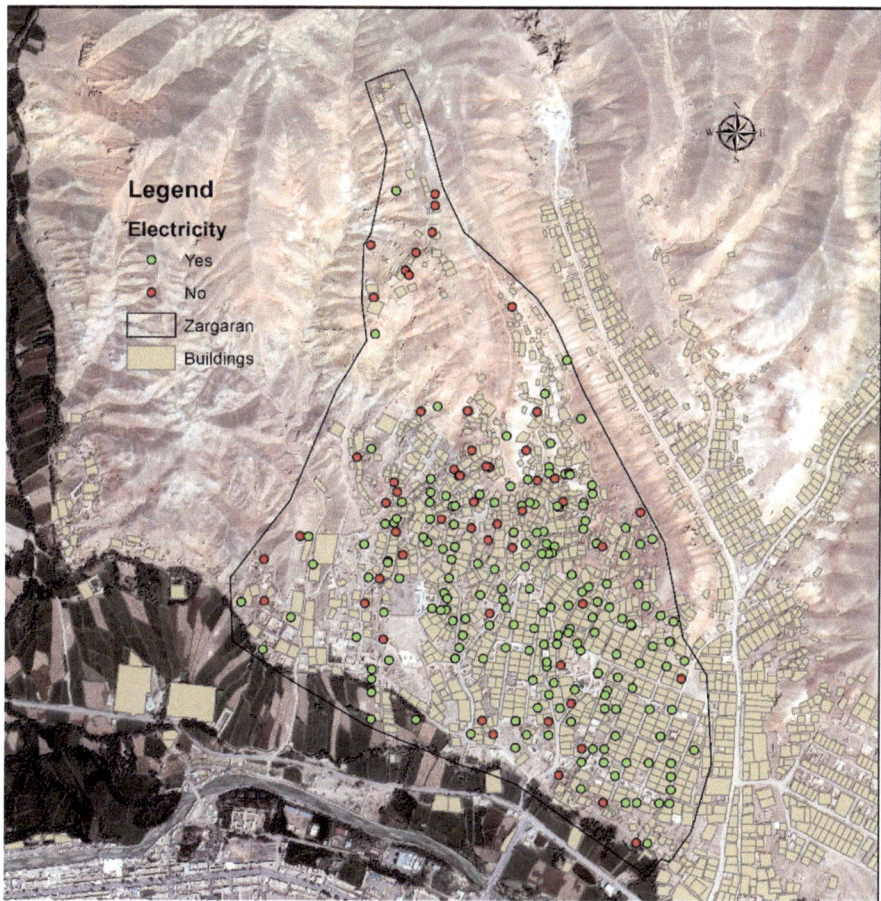

Fig. 7 Electricity (*Source* author's processing of LaGeS Household survey data 2017)

of the village, where grey wastewater simply flows out of the house and into the streets. In addition to hygiene problems, this situation exacerbates the fragility of the built environment. The high levels of moisture in the ground, resulting from the continuous seepage of wastewater, and particularly in areas characterized by high slopes, increases the risk of landslides and mudslides (Fig. 8).

As far as facilities are concerned, there are 39 shops in Zargaran (Ahmadi 2018; UN-Habitat (unpublished)): 25 grocery stores, 6 bakeries, 4 fuel stations and 4 pharmacies. They are located along the two main traffic arteries in the central lower parts of Zargaran, but do not reach the upper part of the village.

In Zargaran there are five community mosques and one state-run elementary school. Furthermore, there are three preschools, either community-run or run by private NGOs.

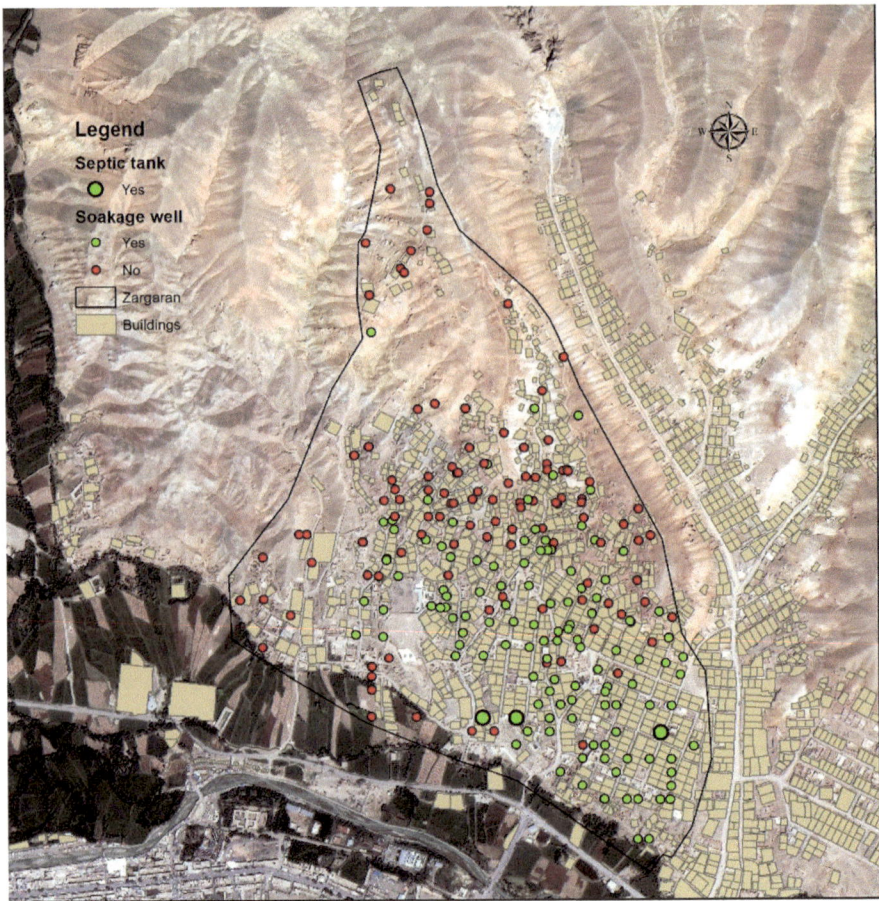

Fig. 8 Sewage cleaning (*Source* author's processing of LaGeS Household survey data 2017)

However, the inhabitants of Zargaran have no access to health services, parks or green areas. There is no nursery, no library, no sports facilities, no police station, no post office or fire brigade in the area.

Solid waste is collected in Zargaran—as in many, but not all, informal settlements in the valley—by the municipality twice a week (Ahmadi 2018).

Finally, 12 Zargaran families dwell in the UNESCO Great Cliff core zone, in breach of the heritage protection regulations, and should be moved to a different place. However, many more families currently dwell in the same core zone, west of Zargaran, which is included in the Dawoodi village. Therefore, addressing this issue would require a major resettlement project (Fig. 9).

Fig. 9 Encroachment in the UNESCO property Nr. 1

3.4 The Road System

Apart from the planned area in Zargaran's south-eastern corner, the road network in Zargaran is clearly insufficient, too narrow and not clearly laid out. The higher parts of Zargraran in particular are hard to reach by car. Taking into account that no street in Zargaran is at the moment paved, the clay soil becomes very slippery and difficult to walk on when wet. However, a UNOPS project is currently being implemented for the reintroduction of plum concrete surface streets in Gozar 1 and 2, for a total length of 573 m (Fig. 10).

Fig. 10 Zargaran's road system (*Source* Ahmadi 2018) and UNOPS Upgrading project

4 Developing the Rehabilitation Plan

The Rehabilitation Plan[7] is proposed as a tool for rehabilitation rather than urban regeneration, in that its aim is to improve livability within the already-existing settlement, rather than to completely alter its essence and its landscape.

While it may be true that Zargaran has infrastructural deficiencies, these difficulties neither clash with territorial balances (see the 2018 Master Plan) nor with existing safeguarding principles (about half of its extension is classified as a buffer zone of the Great Cliff archaeological area), apart from the few buildings within the UNESCO property zone, seen below.

This is largely due to the homogeneity of the buildings throughout the settlement and the settlement's adherence to traditional Afghan building methods. Establishing its layout and physiognomy therefore offers a strategic advantage in attempts to defend the valley from the threat of widespread and damaging structural changes. To properly validate the settlement as a whole, however, many works aimed at infrastructural improvements are needed, ranging from the interior of individual houses to urban networks and large point facilities. In addition, it is the plan's objective

---

[7] According to the Afghan Urban Law (Urban Development and Housing Law, 2016) such a rehabilitation plan would be called Land Resettlement Plan ("refers to a plan based on which, various segments of areas within the cities and suburbs that have witnessed unplanned and non-standard urban expansion and face infrastructural and environmental challenges, will be changed to an organized community and provided with effective urban facilities and equipment"; Article 3, point 16).

to resolve the conflict brought about by the presence of 12 inhabited residential buildings within the UNESCO property zone of the Great Cliff.

To improve the quality of the houses, these operations should include:

- Creating flues for domestic combustion fumes to be funneled outside the house
- Equipping the house with solar panels
- Providing a toilet where it is lacking
- Providing a graywater drainage system.

During winter, residents rely on coal or organic fuel-powered stoves for heating and cooking. Unfortunately, this often results in harmful fumes being trapped indoors, which leads to health hazards. While it may be challenging to completely overhaul heating systems in the short term, installing a flue can help alleviate the negative effects of combustion.

Additionally, installing solar panels can generate electricity for lighting purposes. Some of the energy could potentially be utilized to power a heating system.

On the other hand, recent upgrades to water supply-related facilities have almost eliminated the problem of lack of water in homes entirely.

It is important to pay close attention to the toilet facilities, because in this area toilets are often located outside the house and in external areas.

The waste will be collected in a separate deposit and used for soil fertilization. For new toilets, a septic tank will be constructed and regularly emptied. The grey water will be collected in a channeled system and discharged downstream.

Among the various infrastructural projects to be addressed by the plan, improvements to the current network of route connections is crucial, since it could potentially provide most of the functional, relational, figurative and environmental assets that the settlement lacks today.

In order to maximize effectiveness, it is appropriate to first distinguish between pedestrian and vehicular routes, dedicating the narrowest passages and steepest slopes to the former, compensated by shorter and more direct routes. The latter will instead consist of the widest passages and gentler slopes, even at the cost of longer routes. In this way, the project might facilitate both modes of travel and, in addition, it might offer the inhabitants situated along the pedestrian routes, an open system of relational spaces.

Secondly, regarding road infrastructure, it is seen as necessary to distinguish between the areas with moderate slopes (generally the southern part of the settlement) where straight paths will still be prevalent, and the areas with pronounced slopes, where curvilinear paths will be highlighted instead. This way, it becomes possible to reinforce the different landscapes that characterize the two main parts of the village.

Upgrading Zargaran will involve some street paving to ensure easier connection between the houses and the valley floor. Quarry stones or cement will be used for paving in the densest and steepest areas. In addition, steps and rainwater channels will be installed to improve accessibility.

Upgrading pedestrian routes will make it easier to reach the means of public transport active in the valley, while the presence of basic services in Zargaran will reduce the need for long journeys. The plan will ensure the improvement of the main pedestrian connection systems, compatible with orographic and atmospheric conditions.

Among the facilities in question, there will be certain areas geared towards the provision of basic services, specifically addressed to women and children.

As far as commercial services are concerned, the village already enjoys a fair amount of autonomy due to the presence of 39 stores that cover primary merchandising needs. Moreover, it is located relatively close to the large bazaar at the bottom of the valley, which is the largest commercial resource in the province. Therefore, it is not deemed necessary to explicitly provide for a new commercial cluster in the plan, but rather to merely leave in the management regulations an option for individual businesses to be included.

As for community facilities, a small urban park with a playground will be established, providing opportunities for outdoor family activities.

In addition, it is seen as appropriate to introduce a marginal share of housing completion in Zargaran, which would serve: a. transiently as a shelter for the families whose houses will gradually form the subject of plant upgrading; and b. permanently as housing accommodation for the families relocated from the core zone of the Great Cliff, and for others that need to settle.

We believe that, because of the characteristics of the context to which it applies and because of the goals it sets, which are realistic and at the same time respectful of the local culture, this Rehabilitation Plan may also provide methodological insights useful to other cases of upgrading of informal settlements in different regions and countries.

References

Ahmadi AR (2018) Rehabilitation of informal settlements–Zargaran neighbourhood. Master thesis, Florence University

Djamil B (2016) Sun-dried clay for sustainable constructions. Int J Appl Eng Res 11(6):4628–4633. https://core.ac.uk/download/pdf/47345426.pdf. Last Accessed 04 July 2022

Indexmundi (2023) https://www.indexmundi.com/g/g.aspx?c=af&v=25&l=it. Last Accessed 03 Sept 2023

LaGeS (Laboratorio di Geografia Sociale, Università degli Studi Firenze) (2018) Bamiyan strategic master plan. Polistampa, Firenze

Sruthi GS (2013) Mud architecture. Int J Innov Res Sci Eng Technol 1(2). https://www.rroij.com/open-access/mud-architecture.pdf. Last Accessed 04 July 2022

UN-Habitat (unpublished) Building survey 2019–21

UN-Habitat (2015a) State of Afghan cities, vol 2. Kabul

UN-Habitat (2015b) United Nations settlement programme. Habitat III issue paper 22, Informal Settlements, Nairobi

USAID (2022) Mud brick stabilization for low cost housing, https://reliefweb.int/report/nigeria/mud-brick-stabilization-low-cost-housing-construction-borno-state-exploring-shelter-solutions-idps-living-host-communities-and-returnees-march-2022. Last Accessed 04 July 2022

Improving the Accessibility of Educational Facilities in Bamiyan

Mario Tartaglia⊙ and Masihullah Ahmadzai

Abstract Among the several dimensions involved in the concept of access to education, spatial accessibility plays a leading role in assuring a fair fruition of the school system in a territory. This paper illustrates a specific method developed for evaluating the degree of spatial accessibility of educational sites. Although the proposed method has been developed for the case study of Bamiyan, it can be considered as a general evaluation framework for assessing the accessibility of educational facilities in any populated areas.

Keywords Accessibility · Educational facilities · Spatial analysis

1 Introduction

As stated by several United Nations organizations such as World Education Forum and UNESCO, the access to education is one of the fundamental human rights. The right to benefit from education is also addressed as one of the Sustainable Development Goals (SDGs) set up by United Nations, aimed to ensure inclusive and equitable quality education and promote lifelong learning opportunities for all.

Among the several dimensions involved in the concept of access to education, spatial accessibility is particularly interesting from a geographical point of view. In principle, spatial accessibility refers to the ease of reaching a place or the ease for a place of being reached, and such a dimension is fundamental in the order to guarantee the provision of fair and equal access to educational opportunities.

Accessibility of educational sites has been one of the issues investigated within the projects carried out by the University of Florence team for the area of Bamiyan, Afghanistan, starting from the preparation of the Strategic Masterplan onward

M. Tartaglia (✉) · M. Ahmadzai
Laboratory for Social Geography (LaGeS), University of Florence, Via San Gallo 10, 50129 Firenze, Italy
e-mail: mario.tartaglia@unifi.it

M. Ahmadzai
e-mail: masihullah.ahmadzai@stud.unifi.it

© The Author(s) 2024
M. Loda and P. Abenante (eds.), *Cultural Heritage and Development in Fragile Contexts*, Research for Development, https://doi.org/10.1007/978-3-031-54816-1_11

(LaGeS 2018). The work has been based on several on site surveys, related to spatial and functional land characteristics such as population, settlements, facilities, transport networks. Referring to these projects, this paper shows a further development dealing with spatial accessibility, *id est* an application method aimed to estimate the degree of accessibility to schools.

Although the work is based on the approach proposed by the Ministry of Urban Development and Land of Afghanistan (MUDL) in its Guidelines for Urban Detailed Plans (Urban detailed plan instruction (Dari version) 2019), it can be suitable for assessing the spatial accessibility of a system of educational facilities in a generic spatial context.

2 Theoretical Background

2.1 *Education as a Human Right*

According to United Nations, "human rights are rights inherent to all human beings, regardless of race, sex, nationality, ethnicity, language, religion, or any other status" (United Nations Human Right Webpage 2023). This powerful concept is supported by the International Human Rights Law, a comprehensive body of human rights law which includes—besides several international human rights treaties—the so-called International Bill of Human Rights. The latter in turn is made by both the Universal Declaration of Human Rights (UDHR) (United Nations (General Assembly) 1948) adopted by the United Nations General Assembly in 1948, and the two international treaties that would further shape international human rights: the International Covenant on Economic Social and Cultural Rights (ICESCR) (United Nations (General Assembly) 1966a), and the International Covenant on Civil and Political Rights (ICCPR) (United Nations (General Assembly) 1966b). In addition to many basic rights of people, the ICCPR especially focuses to some rights linked to educations, such as the right to education itself, the freedom of parents to choose schooling for their children, the right to take part in cultural life, the right to enjoy benefits of science, and so on.

Starting from these milestones, the right to education is currently promoted and supported by several governmental bodies and international non-governmental organizations.[1] In particular, it should be mentioned as the position expressed by the World Education Forum[2] in the Dakar Framework for Action (UNESCO 2000), following

---

[1] See for example the initiatives and the primers publications by The Right to Education Initiative (RTE) at https://www.right-to-education.org/.

[2] The World Education Forum is an international body promoted by UNESCO and participated by other international organizations such as World Bank and others. The last conference by World Education Forum has been held in 2023 (see https://www.unesco.org/en/articles/education-world-forum-2023-unesco-mobilizes-ministers-greening-education-and-digital-transformation).

Table 1 The different dimension of the access to educational facilities (*Source* author's processing based on Bruno et al. (2022)

Dimension	Meaning	Spatiality
Availability	Adequacy of the system supply	Spatial
Accessibility	Relationship between the locations of providers and users	Spatial
Accommodation	How providers' resources are organized to satisfy users' demand	Aspatial
Affordability	Users' perception of the costs of the offered services	Aspatial
Acceptability	Users' reaction to personal and practice providers' features	Aspatial
Adaptability	Providers' availability to adapt to individual ability	Aspatial
Horizontality	Characteristic of prestige and quality across the system	Aspatial

the 1990's UNESCO World Conference on Education for All: "Education is a fundamental human right", and "all children, young people and adults have the human right to benefit from an education that will meet their basic learning needs". Moreover, UNESCO recently reiterates that "UNESCO education sector is committed to ensure every child, youth and adult has access to quality education throughout life" (UNESCO 2022). Finally, the definition of education as a fundamental right is confirmed by its inclusion in the United Nations Sustainable Development Goals, as Goal number 4 is stated as "Ensure inclusive and equitable quality education and promote lifelong learning opportunities for all" (United Nations 2023).

For making use of education systems, it is of course necessary to access its facilities. Hence following the international agreement about the education as one of the basic human rights, it is now almost universally accepted that such right has to be ensured by guaranteeing the provision of fair and equal access to educational opportunities is a fundamental right for all citizens (Rekha et al. 2020). However, the concept of access is considerably complex, and it involves different dimensions, only few of which are specified referring to the notion of geographical space (Table 1).

Although some literature items investigate access dimensions like affordability, acceptability and adaptability or appropriateness (see Sharma and Patil 2022, p. 2), most of the studies about access to school are focused on spatial accessibility, as illustrated in the next section.

2.2 Spatial Accessibility to Schools

Even though spatial accessibility has been defined in several ways by different authors, it generally refers to the ease of reaching a place, or the ease for a place of being reached. Starting from the first notable definition given by Hansen as the "potential of opportunities for interaction" (Hansen 1959), the concept of accessibility has been largely investigated in the last fifty years. However, different definitions of spatial accessibility can be found in several research papers, under different points of view, as illustrated for instance in Vannacci et al. (2015). While the majority of the definitions retrieved from literature are focused on one single topic, such as the relevance of destinations (Vandenbulcke et al. 2009) or the structure of land-use (Geurs and Ritsema van Eck 2001), some researcher tried to give a more comprehensive definition of spatial accessibility. One interesting example of extensive definition of accessibility is included in Geurs and Wee (2004), where four different categories of accessibility are identified: infrastructure-based, location-based, person-based and utility-based.

Each point of view in defining accessibility can lead to specific accessibility measures or indicators. Indeed, the main determinants of accessibility are the **land-use** system (the spatial distribution and the characteristics of both destinations and opportunities), the **transport supply** (its structure, the available modes and the levels of service that produce a given impedance in accessing), **time constraints** (availability of opportunities and people willingness to access them), and **people** (people access demand, people needs, abilities and resources of individuals).

All these concepts can be specialized for investigating the accessibility of the educational facilities, so that the land-use is mainly represented by school complexes and residential settlements; transport supply is given by the available transportation networks and services; time constraints are given by educational schedules, transportation service timetables, and people daily organization; and people are represented by the student's population with their needs and abilities.

The notion of spatial accessibility to school systems could be used to focus different issues related to education, such as the equity of a school system in a territory (see some recent examples in Sharma and Patil (2022); Marques et al. (2020); Pizzol et al. (2021)), the improvement or the optimization of the school system (see for instance Armas et al. (2022); Han et al. (2023)), the land planning process (see one example in Sá Marques et al. (2019)), and even sustainability (see Gao et al. (2016)).

Whatever be the point of view of a study about educational spatial accessibility, it needs to have some analytical tool at disposal, aimed to calculate indicators for estimating accessibility levels. For this goal, a very suitable category of techniques is represented by geographic information systems and spatial analysis tools. Indeed, they are more and more used for this kind of studies, especially when fostered by the increasing availability of geospatial information (Park and Goldberg 2021). Interesting examples of application of these techniques for analyzing spatial accessibility to school systems can be found in Rekha et al. (2020); Gao et al. (2016); Aule et al. (2023); Deniz (2023)).

3   Research Design

3.1   The Case Study: The Educational System of Bamiyan, Afghanistan

The city of Bamiyan is characterized by a unique territorial confguration, mainly due to its geographical position in a fertile valley at about 2,500 m above the sea level, surrounded by hundreds of kilometers of mountainous, bare and scarcely populated land. Such location, almost isolated in the middle of Afghanistan, has provided a relative security to the city even during the most dramatic events in the country's history (Loda and Tartaglia 2020).

The mentioned territorial characteristics, together with the socioeconomic and political evolution of the whole Country in the last decades, resulted in a quite poor condition of the urban facilities, including the school system network. Nevertheless, the urban area is provided by all the pre-university educational levels. The General Education framework of Afghanistan comprises three levels: the primary (grades 1–6), the (lower) secondary (grades 7–9) and the higher secondary, or simply 'high' (grades 10–12) (MoE 2015; UNESCO 2015; NUFFIC 2018). Only the primary and secondary levels are compulsory. Moreover, a parallel system of Islamic education, technical and vocational education is available.

In the Bamiyan urban area, 27 different school complexes have been surveyed by the LaGeS team in 2021. They supply more than one education level: the high schools also include primary and secondary schools; the secondary schools also include the primary school level. As shown in Fig. 1, the Bamiyan school facilities are almost fairly distributed within the municipality area, even if different levels of education are provided in different locations of the city.

3.2   Methods and Data

Il metodo di analisi sviluppato nel presente studio è basato sull'utilizzo di funzioni di analisi spaziale normalmente disponibili in un generico Geographic Informa-tion System. È stata inoltre posta particolare attenzione nell'assicurare la compati-bilità della procedura di calcolo con l'approccio normativo dettato dal competente ministero del governo Afghano ed in particolare con the Guidelines for Urban Detailed Plans released by the Ministry of Urban Development and Land of the Islamic Republic of Afghanistan (MUDL), that contain some accessibility rules for educational facilities (Table 2).

However, the mentioned guidelines have got some haziness. They don't specify any distance measurement methods so that it is unclear which type of distancing system they refer to, e.g. Eulerian, network constrained, or others. Moreover, the maximum accessibility radius and the target population values are defined as a range, giving room for a large number of different possible choices in their usage for the

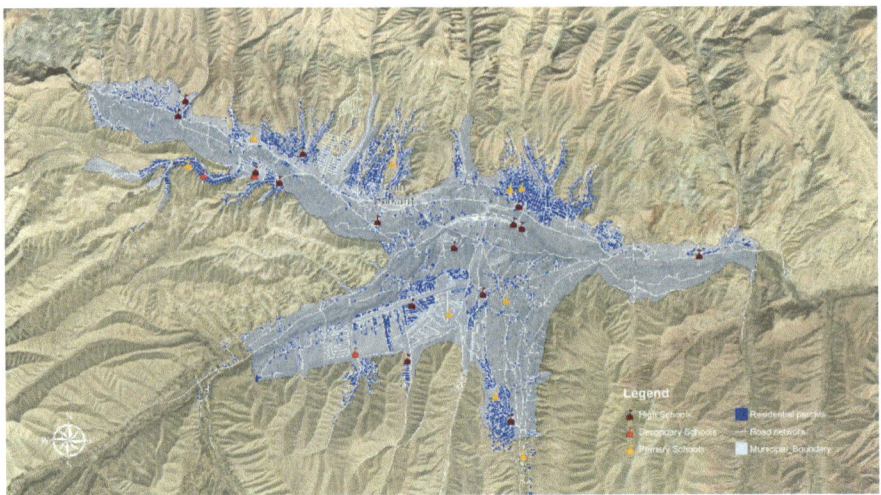

Fig. 1 Spatial distribution and type of the Bamiyan's school complexes (*Source* author's processing of LaGeS data)

Table 2 Accessibility rules for educational facilities (*Source* Urban detailed plan instruction (Dari version) (2019))

Type of facility	Max accessibility radius (m)	Walking time (min)	Target population (inh.)
Kindergarten	300–500	5 ÷ 10	1000–2500
Primary school	500–800	10	5000–15000
Secondary school	1000–2000	15	10000–50000
High school	2000	20	10000–60000

analysis. In addition, association between the proposed walking times and distances leads to speeds between 3 and 8 km/h. These does not fully comply to the pedestrian average walking speed that it's generally considered around 5 km/h, and the criteria of their variation is unclear as well.

Come già accennato sopra, i dati sul sistema scolastico sono stati direttamente rilevati dal team di ricerca attraverso survey in situ, includendo informazioni sulla posizione geografica, the education level provided, the school ownership, the gender allowed, the number of shift, the average number of students and more. According to these data, among the 27 school complex surveyed, only 16 are high schools providing all the three educational levels, 3 are secondary including the primary level, 8 are just primary schools and no kindergartens have been found during the survey. Nineteen schools over 27 are public and most of the schools are open to both genders; only 6 are only for female and 3 only for male students.

Oltre ai dati relative alle scuole, è stato ritenuto necessario includere nell'analisi due altri set informativi chiave rilevati in situ dal team di ricerca. Il primo è dato dall'entità e dalla distribuzione geografica della popolazione dettagliata per residential parcel nell'area in esame. Il secondo è costituito dalla struttura della rete stradale, specificata attraverso la sua geometry, topology and plano-altimetric characteristics, including the slope of each road section. Mentre il primo set di dati è utile a rappresentare l'entità della domanda di educazione, il secondo consente di calcolare la difficoltà di accesso ai diversi siti.

La metodologia messa a punto è descritta nella Fig. 2 nella quale si vede che a partire dai dataset sopra menzionati, l'accessibilità di ogni school complex viene modellizzata utilizzando funzioni dell'analisi spaziale finalizzate a costruire both the iso-distance geographic shapes from which it is possible to get to the educational facilities from different distances, and a potential accessibility indicator able to take into account also the educational demand.

According to this framework, two different measures for estimating accessibility to schools have been adopted in the present study:

1. the **accessibility distance**, *id est* the distance that students living in the study area are expected to travel, on average, to reach the closest school.
2. the **potential accessibility**, *id est* the percentage (or the number) of students in the study region having the closest school within given distance from their residence site.

Since 80% of daily trips are made on foot, and also considering that this percentage should be higher for the school-aged population, the distance bands considered in the calculation have been thought compatible with walking. Moreover, the potential

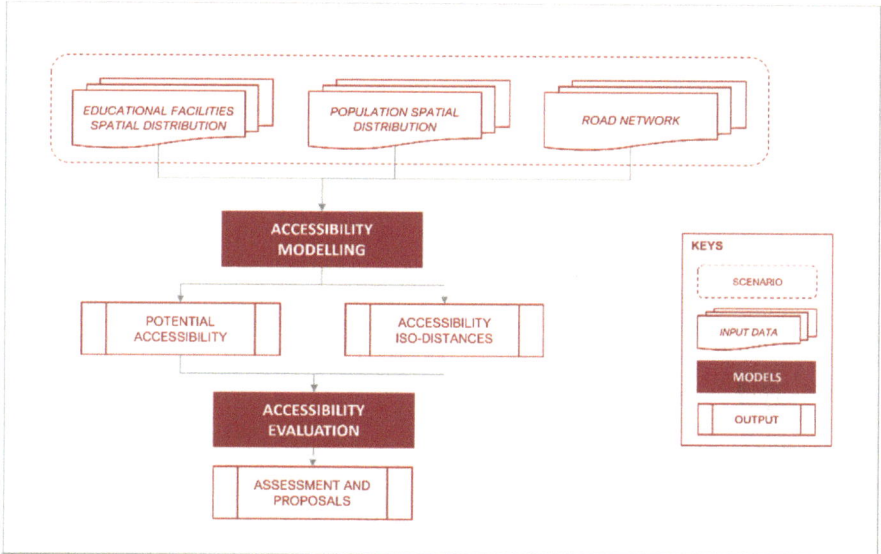

Fig. 2 Diagram of the developed methodology for assessing the spatial accessibility level to educational system (*Source* author's own processing)

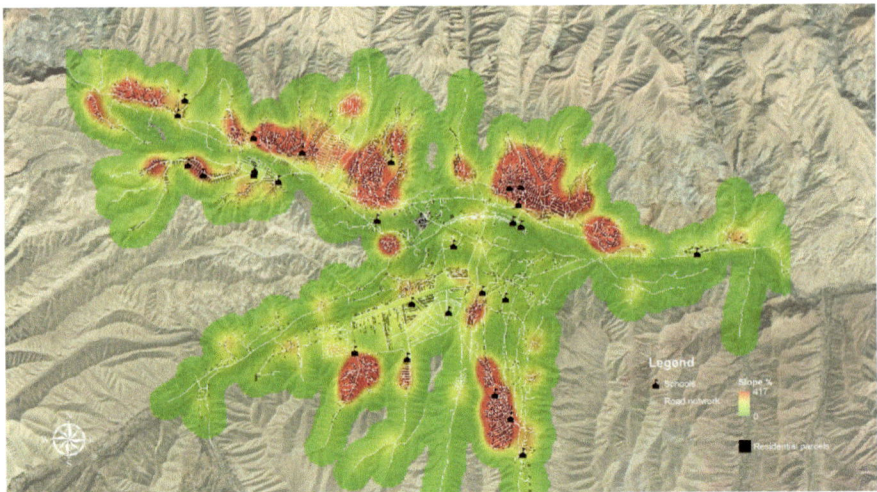

Fig. 3 Spatial distribution of road average slopes in Bamiyan (*Source* author's processing of LaGeS data)

accessibility indicator is calculated as the number of people belonging to the age class corresponding to the educational level that is resident within a given iso-distance geographical shape.

The road slope has been estimated by assigning the average slope of the terrain to each road section, using a suitable open source Digital Terrain Model. The need of this additional calculation derives from the mountainous nature of the considered land; indeed, the effect of the slope is clearly visible in Fig. 3, where the red coloured higher grades are concentrated on the mount sides apart from the valley axes, especially in the north. After determining the iso-distance shapes and the potential accessibility indicators for each school or school level, it is possible to develop an evaluation of the accessibility level and hence assessing it against the chosen standards or guidelines. Such benchmark analysis finally leads to the development of proposals or projects aimed to improve the general level of service of the educational system.

4   Results and Discussion

The application of the illustrated methodology first led to build several sets of iso-distance geographic shapes around each school complex. The overall picture of the calculated iso-distance areas around the schools in Bamiyan isshown in Fig. 4, where one can see that the populated land is broadly included within a walking distance of 2 km, with the exception of the mount sides in the North-East (village of Jugra Khail) and the valley the runs towards South-West. Anyway, all the central part of the

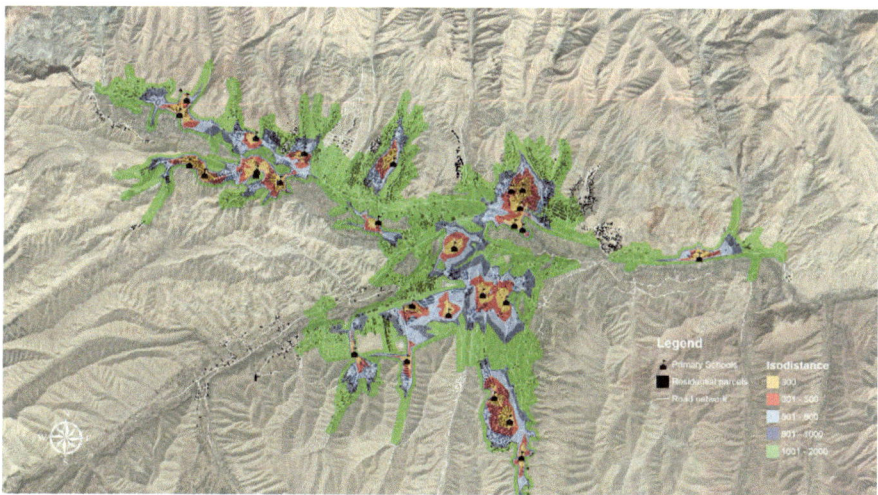

Fig. 4 Iso-distance areas around the school complexes in Bamiyan (*Source* author's processing of LaGeS data)

urban area is included in the 2 km iso-distance shape, as well as the main West–East valley, the southern valleys and the cliffs limiting the city in the North.

A different picture emerges considering only the accessibility areas to the primary schools as defined by the MUDL guidelines, i.e. limiting the walking distance to 500 m. In this case, as shown in Fig. 5, most of the settlement are excluded from the accessibility areas, due to the short accessibility radius adopted and in spite of the fact that all the school complex hosts primary teaching. As regards to the gender of allowed pupils, about the same surface is available for boys and girls.

The accessibility areas for secondary schools, illustrated in Fig. 6, cover a larger portion of the land, including more settlement units; although the number of school complex is smaller, thanks to a higher accessibility radius considered. It's to be noted that most of the northern faces are excluded as well as the northern side of the South-West valley. Also in this case, the amount of settlements included in the accessibility areas available for boys is slightly greater than the one available for girls.

Finally, adopting a maximum walking distance of 2,000 m, the overall accessibility area associated to the high schools cover most of the urban land, although the high school complexes are fewer (Fig. 7). An unsupplied area stands out in the upper western sector to the North, corresponding to the villages of Sang-e Chaspan, Khwaja Kamalodin and Surkh Qol, where only a primary school site is present. Another unsupplied cluster of parcels is located in the South of Dasht-e Esa Khan new development zone. As for the lower school categories, the accessibility area of the schools available for male pupils is slightly larger than the one available for girls.

Even if it shows an interesting insight about the accessibility areas to the schools the situation depicted by Figs. 4, 5, 6 and 7 remains purely geometric, i.e. based only on the supply facilities available, as far as one doesn't consider the real distribution

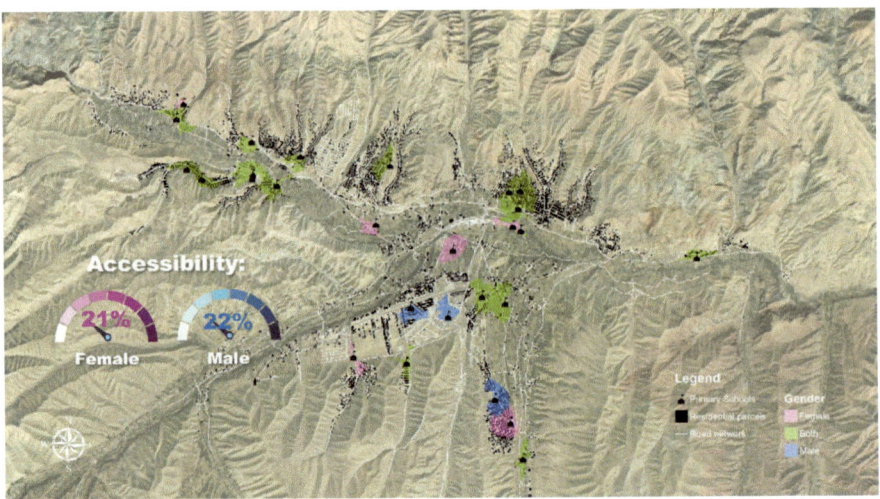

Fig. 5 Accessibility areas for primary schools in Bamiyan, according to the 500 m MUDL accessibility radius (*Source* author's processing of LaGeS data)

Fig. 6 Accessibility areas for secondary schools in Bamiyan, according to the 1,000 m MUDL accessibility radius (*Source* author's processing of LaGeS data)

of population on the land. For this reason, our analysis went forward and took into account also the population distribution by age classes. In the absence of detailed data about this item, such variable has been calculated starting from the distribution of the inhabitants over the residential parcels by means of the same demographic model implemented for the Bamiyan Masterplan (LaGeS 2018). The population

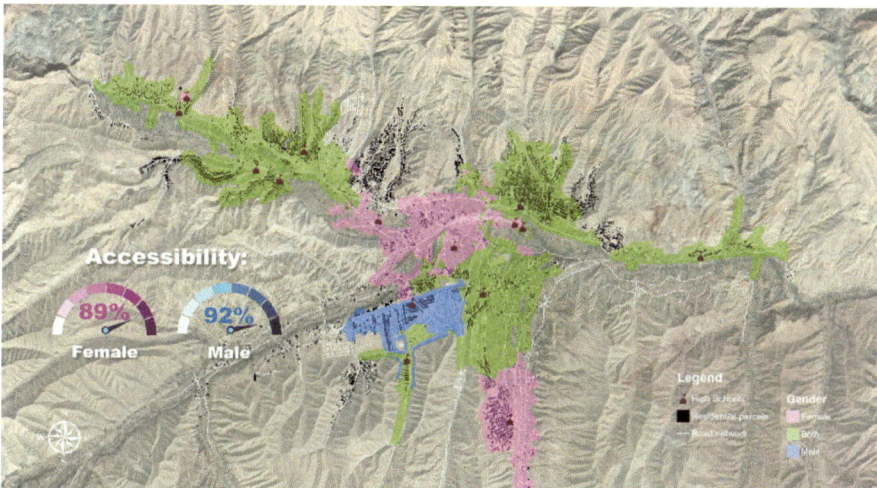

Fig. 7 Accessibility areas for high schools in Bamiyan, according to the 2,000 m MUDL accessibility radius (*Source* author's processing of LaGeS data)

distribution by age classes well represents the real educational demand since each age class corresponds to an educational level among primary, secondary and high. Hence, a significant indicator of the educational spatial accessibility classified by level is the number of people belonging to the age class matching a given educational level included in the corresponding accessibility area. Such an indicator is shown in Fig. 8, from which it is evident that the potential accessibility pattern looks like a negative exponential curve as a function of distance, according to literature. Moreover, it should be highlighted that the average potential accessibility to schools is rather low for the shortest distance catchment areas (e.g. 50% student population located within 1 km), while it is quite high (90%) if we consider catchment areas extended to within 2 km walking distance. In general, in over about 20,000 people in the school age range as calculated in this study, we found that about 18,000 are resident within a 2 km accessibility radius from the schools.

However, with reference to the MUDL guidelines about spatial accessibility, only high school students result to be supplied with a reasonable level of service, being 90% within the admitted range. Quite the opposite, the secondary and primary school students are only 50% and 20% within the accessibility range respectively.

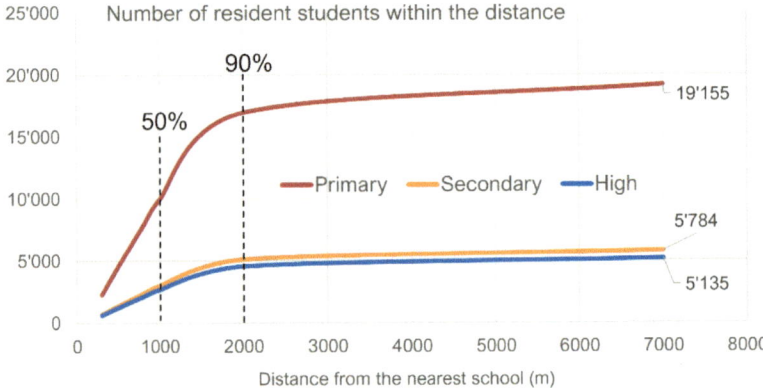

Fig. 8 Potential accessibility of Bamiyan schools by type (*Source* author's processing of LaGeS data)

5 Conclusions

An accessibility assessment method based on spatial analysis techniques has been developed and applied to the case of the school system of Bamiyan, Afghanistan, in the order to check the agreement to the requirements expressed by the MUDL guidelines about urban development plans.

The application to the Bamiyan case showed that the compliance to the MUDL spatial accessibility requirements is poor for primary schools (slightly more than 20% students in the 500 m catchment area); not satisfactory for secondary schools (around 50% students in the 1000 m catchment area); quite good for high schools (around 90% students in the 2000 m catchment area). However, the school spatial availability is slightly lower (1 ÷ 3% percent) for the female student population. Moreover, the calculation showed that some urban zones are clearly disadvantaged as regards the spatial accessibility to schools, due to long distances to be walked, with differences according to the different school grades. Some other urban zones are disadvantaged due to terrain slope along the road network.

These results allow to develop some proposals for improving accessibility to the Bamiyan school system as regards the spatial accessibility. First, as the school specialization by gender lower the accessibility level, the number of schools available for both genders should be increased. Moreover, concerning primary and secondary schools, as a large part of urban settlement is out of the MUDL accessibility range, and several schools are clustered in specific areas, the spatial distribution of schools should be optimized, in the order to increase the accessibility of some urban zones (South-West Valley, Karak Valley, Petab-e-Laghman and Mullayan to the East, the middle of the city, Jugra Khail close to Zargaran). About the high school situation, the accessibility to schools should be improved in a few urban zones: Surkh Qhol and Sang-e-Chaspan to the North, Petab-e-Laghman to the East, Azhdar to the West, and the South-West Valley). All these proposals could be obtained by enhancing the transport system or increasing the number of school complexes, or even by a combination of these two measures.

Finally, the method aimed at the assessment of spatial accessibility to school systems illustrated in this paper can be considered a useful tool for assessing the spatial accessibility of a system of educational facilities in a generic spatial context.

CRediT authorship contribution statement. Mario Tartaglia: Conceptualization, Methodology, Supervision, Formal Analysis, Writing—Original Draft, Visualization, Writing, Review & Editing. Masihullah Ahmadzai: Data curation, Formal Analysis.

References

Aule DS, Jibril MS, Adewuyi TO (2023) Geostatistical analysis of accessibility to secondary schools in parts of Benue State. J Geogr Inf Syst 15:1–18

Bruno G, Cavola M, Diglio A et al (2022) Geographical accessibility to upper secondary education: an Italian regional case study. Ann Reg Sci 69:511–536

de Armas J, Ramalhinho H, Reynal-Querol M (2022) Improving the accessibility to public schools in urban areas of developing countries through a location model and an analytical framework. PLoS ONE 17(1)

Deniz M (2023) Analysis of accessibility to public schools with GIS: a case study of Salihli city (Turkey). Children's Geographies

Gao Y, He Q, Liu Y, Zhang L, Wang H, Cai E (2016) Imbalance in spatial accessibility to primary and secondary schools in China: guidance for education sustainability. Sustainability 8(12):1236

Geurs KT, Ritsema van Eck JR (2001) Accessibility measures: review and applications. Evaluation of accessibility impacts of land-use transportation scenarios, and related social and economic impact. Netherlands Environmental Assessment Agency, Technical report 90715

Geurs KT, van Wee B (2004) Accessibility evaluation of landuse and transport strategies: review and research directions. J Transp Geogr 12:127–140

Han Z, Cui C, Kong Y, Li Q, Chen Y, Chen X (2023) Improving educational equity by maximizing service coverage in rural Changyuan, China: an evaluation-optimization-validation framework based on spatial accessibility to schools. Appl Geogr 152:102891

Hansen WG (1959) How accessibility shapes land use. J Am Inst Plan 25(1):73–76

LaGeS (2018) Laboratorio di Geografia Sociale: Bamiyan strategic masterplan. Polistampa, Florence

Loda M, Tartaglia M (2020) Developing the Bamiyan master plan between cooperation and politics. Rivista Geografica Italiana CXXVII(2):29–49

Marques JL, Wolf J, Feitosa F (2020) Accessibility to primary schools in Portugal: a case of spatial inequity? Reg Sci Policy Pract 13(3):693–707

Ministry of Urban Development and Land (2019) Urban detailed plan instruction (Dari version), pp 49–50. Islamic Republic of Afghanistan

MoE (2015) Ministry of Education, Islamic Republic of Afghanistan: Afghanistan National Education for All (EFA). Review Report, Kabul

NUFFIC (2018) Dutch organisation for internationalisation in education, education system Afghanistan

Park J, Goldberg DW (2021) A review of recent spatial accessibility studies that benefitted from advanced geospatial information: multimodal transportation and spatiotemporal disaggregation. ISPRS Int J Geo-Inf 10:532

Pizzol B, Giannotti M, Tomasiello DB (2021) Qualifying accessibility to education to investigate spatial equity. J Transp Geogr 96:1–11

Rekha RS, Radhakrishnan N, Mathew S (2020) Spatial accessibility analysis of schools using geospatial techniques. Spat Inf Res 28:699–708

Sá Marques T, Saraiva M, Ribeiro D, Amante A, Silva D, Melo P (2019) Accessibility to services of general interest in polycentric urban system planning: the case of Portugal. European Planning Studies

Sharma G, Patil GR (2022) Spatial and social inequities for educational services accessibility-a case study for schools in Greater Mumbai. Cities 122

UNESCO (2000) Dakar framework for action: education for all. Meeting our collective commitments. In: World forum on education, Dakar, Senegal, 26–28 April 2000. UNESCO, Paris

UNESCO (2015) Education for All (EFA), National review report Afghanistan

UNESCO (2022) Transforming education for the future. UNESCO, Paris

United Nations (General Assembly) (1948) Universal Declaration of Human Rights

United Nations (General Assembly) (1966a) International covenant on economic, social, and cultural rights. Treaty Ser 999

United Nations (General Assembly) (1966b) International covenant on civil and political rights. Treaty Ser 999

United Nations (2023) Sustainable development goals webpage, https://www.un.org/sustainabledevelopment/sustainable-development-goals/. Last Accessed 25 Aug 2023

United Nations Human Right Webpage (2023) https://www.un.org/en/global-issues/human-rights. Last Accessed 25 Aug 2023

Vandenbulcke G, Steenberghen T, Thomas I (2009) Mapping accessibility in Belgium: a tool for land-use and transport planning? J Transp Geogr 17:39–53

Vannacci L, Tartaglia M, Navajas E, Rotoli F (2015) The use of Open Data for estimating rail accessibility in Europe. Ing Ferrov 7–8:611–634

National Museum of Aleppo

Marina Pucci⊕

Abstract This paper presents the history, current situation, and ongoing projects on the collection from the National Museum of Aleppo (Syria), sketching the history of its collection and the changing role of the museums in the past century. When museums face periods of conflict and destruction, not only their primary role to preserve heritage is challenged, but also the bond to the local community is severed. The National Museum of Aleppo is a perfect example in this matter: established under the French protectorate, rearranged by the Syrian government at the end of the 60s, it experienced a flourishing period, the violent conflict, a partial reopening, and the earthquake. Its collection, that was never looted, thanks to the efforts of the museum staff and the DGAM, has been damaged by shelling, has been in great part taken out from showcases, brought to safe deposits, and still not available to the public. Given this situation intervention to the collection and to the structure needs to embed the recent events and aims at reinstalling a new bond between the community and its cultural heritage.

Keywords Museum studies · Syrian archaeology · Conflict areas

1 Background

1.1 Museums' Role and Definition

In Western cultures, museums began to exist in the seventeenth century CE based on private collections of ancient artefacts, they enlarged and increased in the seventeenth–eighteenth centuries. Their first definition as "Wunderkammer" or public

My warmest thank goes to the museum staff members, Ahmed Othman and Desbina Baslan in Aleppo and to Houmam Saad (vice-director of the DGAM in Damascus). Their dedication to the Syrian cultural heritage goes well beyond their professional duties.

M. Pucci (✉)
Dipartimento SAGAS, Università degli Studi di Firenze, via San Gallo 10, Firenze, Italy
e-mail: marina.pucci@unifi.it

© The Author(s) 2024
M. Loda and P. Abenante (eds.), *Cultural Heritage and Development in Fragile Contexts*, Research for Development, https://doi.org/10.1007/978-3-031-54816-1_12

collections suggests that they were strongly related to the individual criteria, interests and knowledge of a single person, family or royal institution that began the collection and at a given point decided to open it to the public in order to show their "mirabilia" (Abt 2006). The main aim of the first large museums during the nineteenth century, inheriting it from the previous experience, was to instill a feeling of wonder, as the German definition suggests, in the audience given by the exotic, out of the ordinary, and foreign features of the artefacts. Even if the concept of collections goes beyond the western world, the public museum seems to be a western product exported during the colonial period (Varutti 2014), i.e. in different moments, both in western and in eastern Asia.

Its definition changed from the nineteenth century, from being in 1946 "all collections open to the public", to "a non-profit making, permanent institution in the service of the society and its development, and open to the public".

In Prague, on 24 August 2022, the Extraordinary General Assembly of ICOM has approved the proposal for the new museum definition: "A museum is a not-for-profit, permanent institution in the service of society that researches, collects, conserves, interprets and exhibits tangible and intangible heritage. Open to the public, accessible and inclusive, museums foster diversity and sustainability. They operate and communicate ethically, professionally and with the participation of communities, offering varied experiences for education, enjoyment, reflection, and knowledge sharing" (ICOM 2023). Thus, it is not a one-way communication (from the institution to the public) but a dialogue that uses the exposed objects to convey exchange. It is also obvious that museums fulfil their function collecting, interpreting and exhibiting "heritage": regardless of the ownership of this heritage (Hodder 2010), the material culture preserved in the museums is the reason why these dialogue spaces exist.

1.2 Museums in Conflict Areas

Numerous events of looting and destruction of museum buildings and their collections over the last hundred years have shown how much the museum has in fact become a non-secondary victim of war damages. Museum buildings are the target of terrorist attacks and of intentional plundering and destruction. They suffer the bombing of the city, the lack of control, and the consequent political instability provided by conflict. Being a public building that hosts materials considered valuable also from an economic point of view, a museum located in a conflict area is the target for antiquity dealers and common thefts; national museums, being a symbol of the cultural heritage of a country, can also become the target for intentional destruction and cultural cleansing. Examples from Afghanistan, i.e. Kabul Museum during the Soviet-Afghan war in the 90s (Stein 2015) or from Baghdad (Gibson et al. 2008) are the most famous ones, however, this phenomenon is widespread and common and it affects all war zones and all museums from national to small local ones (i.e. the 2023 looting of the Sudan National Museum in Kartum). Since 1954, with the Secretary-General's Bulletin on Observance by United Nations Forces of

International Humanitarian Law, the UN force is prohibited on attacking cultural heritage sites, including museums (Kalshoven 2005), according to the New Policy on Cultural Heritage (2001), the concept that pillaging and destruction of cultural heritage, including the museums, are war crimes, has been reinforced, setting them in connection with the crimes against humanity. These statements do not prevent looting and destruction, but at least provide a legal frame of intervention during and after the conflict in order to restore collections and "re-exhibit" the artefacts. It is however true that all post-conflict interventions carried out in museums should decide if they aim to erase the looting actions (restore) or embed this in a new story telling.

In addition, other actions should be taken into consideration, not because, as the ones described above, destroy the cultural heritage, but because they disrupt and somehow affect the role of the museum during and after the conflict. All measures that try to prevent museum looting and destruction are somehow interrupting the role of the museum as a space for interaction and exchange: museum are preventively closed, reinforced, fortified; artefacts are removed from showcases, packed in boxes and sometimes removed from the museum in order to be stored in safer places; large artefacts are covered, hidden, protected with sandbags, wooden frames and whatever materials are easy to be found; museum buildings are used as shelters, installations in the museum are reemployed for everyday use. All these actions, that are carried out for the good and the preservation of the material heritage, still, when they are extended over a long time, affect the museum's role and its museography in a post-conflict situation.

2   Museums, Cultural Heritage, and Identity in Syria

The earliest scientific archaeological research projects in the Syrian region and specifically on the current border between Syria and Turkey were the German excavations located at Zincirli (1888) and Tell Halaf (1899), and in the English excavations at Karkemish (1876),[1] with the well-known aim to deliver many important "pieces" for museum collections in Berlin (Tell Halaf Museum, Vorderasiatisches Museum) and London (British Museum).

After the First World War with the Sykes-Picot agreement (1916) and the Sévres agreement (1920) both in place, Syria came under a French control. In 1920, Chamonard explains that scholarly opinion felt that local museums were needed in the country to create a bond between the local population and their antiquities. A bond, which the French saw as completely absent blaming looting and illicit digging on this ignorance (Chamonard 1920). They saw the creation of two centralized museums and a third regional one in the "new" country as an impelling need: calling for one Museum in Beirut dedicated to ancient art, one in Damascus dealing

---

[1] Although two of the three sites are located in Turkey just on the other side of the border, they are considered geographically strongly related to the Syrian cultural sphere (Wartke 2005).

with Classical and Arabic art and a third regional one in Adana. In 1920, the museum in Beirut consisted of sixty pieces located in a hall of the Prussian deaconess. The museum in Damascus had its starting core in the Arab Academy at Damascus, which was founded in 1919, and first homed in the madrasa al-'Adiliya. It started to collect art objects during the same year, choosing objects "which made Syria great" and were borrowed or donated by wealth "patriotic" Syrian families. It was only in 1924 that this museum of the Arab Academy at Damascus was called the National Museum of Damascus, and was moved to a building near the Sultan Sulayman Mosque in 1936 (Rossi 2011; Virolleaud 1924).

During this period, the journal "Syria" (1920) was founded with the aim of promoting the art and antiquities of Syria and publicize Syrian art from every period. A new antiquity law was drafted (1920) and issued (1923). A Service des Antiquitès et Beaux Arts was constituted (1919) for both Syria and Lebanon (Pottier et al. 1920). French cultural politics during its mandate on the one hand promoted what they thought it was a process of acculturation, the *"mise en valeur"* of the cultural heritage (Ouahes 2018), on the other fostered the nationalistic idea of using cultural heritage as a mean to reinforce the identity of a nation that did not exist before. Several small local archaeological museums were opened during this period, e.g., Lattakia (1929) and Tartus (quoted in 1930).

The French cultural policy and the administrative infrastructure they built in the country had a strong impact on the country also after the constitution of the independent Arab Republic of Syria (1949): the *Direction Générale des Antiquités* became *la Direction Générale des Antiquités et des Musées (DGAM)* in 1951, and a new Antiquities Law was drafted in 1963 (Gillot 2010). The two main national museums were enlarged: new wings were added in 1956 and 1975 to the Damascus archaeological museum. Hama (1956) in the 'Azm palace, Deir ez-Zor (1974 renewed in 1996), al-Raqqa (1981), and Idlib (1989). Most included, besides the pre-Classical and Classical collections, ethnographic sections, which proclaimed their aim of bringing archaeological data and the region's recent history closer together (Maqdissi 2008) basically following the same program started by the French, i.e. teaching the communities to accept and protect their archaeological heritage as part of the "Syrian" tradition and culture.

In the catalog of the Canadian exhibition in 1993, Najah al Attar (Syrian Minister of Culture from 1976 to 2000), highlighted the remarkable growth of fieldwork in the country and the "development" of archaeology in Syrian universities, both of which would potentially produce a better awareness and understanding of Syrian cultural heritage (Cluzan et al. 1993). She expressed the same sentiment three years later in the introduction to the catalog of the exhibition organized by the European Community and the Archaeological Museum in Damascus: joint archaeological projects should be supported in search for common cultural roots (Museum 1996). In 1999, the Minister of Culture remained on the same path, attributing Syria's archaeological remains as an example of world heritage, again emphasizing the role played by Arab countries in the birth of civilization and in its expansion to the West (Fortin 1999). According to her, Syria exported and co-organized these sorts of exhibitions to bridge the gap between different worlds and to promote understanding. While there was no declaration connecting cultural heritage with national identity, abandoning the path traced by the French, all statements seemed to refer to a common "Arab" identity, of which Syria was a part. Thus, although the Syrian government has been supporting local archaeological research since the 1960s, welcoming new excavations, building new museums, adding ethnographic section to archaeological museums, it did not explicitly use for a nationalist agenda, as had happened in Iraq, cf. (Bernhardsson 2005).

Only during the recent conflict in Syria and the consequent intentional destruction of artefacts and sites, the government begun to describe local cultural heritage as belonging to Syrians alone, representing the Syrian collective memory and its multifaceted history.[2] In a conflict situation as the Syrian one, the role of cultural heritage shifted from being the representation of the Arab world, to symbolize the nation in an effort to keep the country united and present the government as the protector of the cultural heritage against other groups that aimed at destroying it.

3   The National Museum of Aleppo

3.1   From a Storage Facility to the "National Museum"

Based on the Sévres agreement, the French planned to include Cilicia in their Mandate of Syria, and the construction of a regional Museum in Adana was planned as part of the cultural politics for governing Syria: it was intended to house the pieces collected by the Colonel Normand and the French army in Cilicia (Normand 1921). Because of the French agreement with the Young Turks (1921) and the reestablishment of the Turkish border just south of the Taurus and Amanus mountains (1922), this project was never accomplished. Only then the French mandate decided to design a third museum (after Damascus and Beirut) in Aleppo. All excavations in northern Syria such as Tell Ahmar, Ugarit, Tell Halaf and Hama started to send their pieces to Aleppo.

Until 1925 these pieces were stored in the rooms of a local boys' high school with no public access. In 1926 the Archaeological Museum of Aleppo was officially founded and located in the Djemilieh structure (1926–1930), but due to lack of space it did not open to the public, so basically it was not a museum but rather a storage facility for the archaeological excavations. In 1924 the al-Naoura palace was built where large gardens were planted directly on the dried bank of the Queiq river. The building was supposed to host the meeting of the House of Representatives (Fakhro 2020), became the residence of general Gaston-Henri Billotte until 1926. This building was chosen to host the collections in 1931 and was soon opened to the public and accessible until 1959, when the collections were temporarily moved in order to allow the new construction of museum where it is still located (Dussaud

---

[2] Abdulkarim in an interview published in Borghese (2015). Cf. also (Qassar 2015).

1925, 1931; Ploix de Rotrou 1932). The necessity to have a larger structure to host the ever-growing collection of the museum was only one of the factors that brought to the decision to dismantle the Ottoman building and replace it with a new modern structure, in addition to that the fact of dismantling a residence symbolizing the French colonial period and a monument to a French general was an integral part of Syrian internal politics at the end of the 50s (Fakhro 2020).

The new modernist structure was built according to the winner plan of an international contest carried out in 1956/7; its construction begun in 1961 and the plan included large underground storage rooms dug directly into the former riverbed as well as a second building for the offices of the antiquity department of the DGAM. The Museum opened in 1969, its museography followed both a topographic and chronological order, grouping artefacts and installation according to the site and arranging the halls following an approximately chronological order. It basically reproduced the same cultural-historic concept that was introduced in the first arrangement (Ploix de Rotrou 1932) but including many more artefacts gathered from excavations over the years. The main issue that raised already during the construction of the new building was the presence of water in the underground rooms, clearly coming from underground spring related to the Queiq river. This problem was controlled through the installation of three water pumps under the floor of the underground facility that would prevent the flooding of the storage.

The building upheld its role throughout the whole twentieth century, ranking as one of the most important museums in Syria and representing a key building for the city of Aleppo, where not only the second largest collection of pre-classical artefacts in Syria was kept and displayed, but also serving as a vital structure able to attract tourists, house conferences and temporary exhibits. Its large front courtyard open to the public was—and, in part, still is—an open park with sculptures where Aleppinians and tourists could stop and rest, and its collection clearly mirrored a feature typical of the city of Aleppo: a place where different religious and ethnic groups lived together, building a unique and multifaceted community. Although visitors' numbers for the museum are not available, the museum's visit was one of the usual stops in every tourist route in a touristic market that registered a 500% increase in the number of visitors from 1999 to 2011 (CEIC Syria Visitor Arrivals 2023).

3.2   The Museum During the Conflict

The museum closed to the public in 2011/12; in October 2012 with the ongoing conflict in Aleppo and the Saad Allah square bomb attack that damaged its windows and showcases, the DGAM decided to remove all small artefacts from the exhibit and store them in a safer place, first inside the museum itself, then at the University of Aleppo (2013) and, later, when the situation in Aleppo was worsening, in the National Museum in Damascus. These operations were carried out with the help of

Aleppinian volunteers on the first stages of packing[3] and with the invaluable help of the museum staff during the whole process: they risked their lives to bring the artefacts to Damascus and to control the museum itself. At the same time the country and the city were experiencing battles and conflict on the ground, power and water shortages,[4] absence of food and provisions. From 2013 to 2019 the museum remained a closed building, accessible only to the staff on what had become a frontline between two sectors of the city controlled by opposing groups.

The museum collection still hosted the large, monumental artefacts that could not be transported to Damascus, such as large statues, architectural and funerary installations that were provided with provisional protection from shelling and bombing (Abdulkarim and Cunliffe 1954; Fakhro 2021), so that no monuments were visible from outside, and even its iconic entrance with copies of monumental statues from Tell Halaf was protected with sandbags and completely sealed. During this period no connections between the museum and the local community existed, and it kept its function of protecting the archaeological heritage that was left in it, but its social function was completely absent (Fig. 1).

4 The Post Conflict Scenario

4.1 The Torn City

After the conflict ended in Aleppo in 2016, the city was heavily damaged, with 30% of the old town completely destroyed (UNESCO 2023) and the population changed: refugees and exchanged prisoners from all neighboring villages fled into the town, large groups of Aleppinians left the city during the conflict. These movements caused a profound disruption in the social composition of the local community.[5] In fact not only Aleppo population shrunk to approximately 35% of its population in 2011 (Munawar and Symonds 2022), but its composition also changed: artisans, traders, shop owners moved their affairs and their lives in areas where it was still possible to work, being replaced by refugees seeking protection in the city. Most of these refugees stayed in Aleppo either because their hometown and villages were and are still under the control of other groups (as the Afrin valley) or because their villages lack any infrastructure and suffered major damages. These are the new Aleppinians

---

[3] The first packing activities carried out by local community members with the local museum staff gave to all participants a sense of community and involvement in protecting the heritage. The documentary movie "The Oath of Cyriac" includes several interviews with the museum staff who worked during these years for the safeguard of the collection.

[4] During the water shortages that affected the city several inhabitants used the water that kept flooding the underground spaces of the building (personal communication D. Baslan).

[5] Many damage assessment reports are available on the city of Aleppo, cf. (World Bank Group 2017) also using remote documentation. Most recent publication on social composition and world heritage is presented in Bandarin et al. (2022).

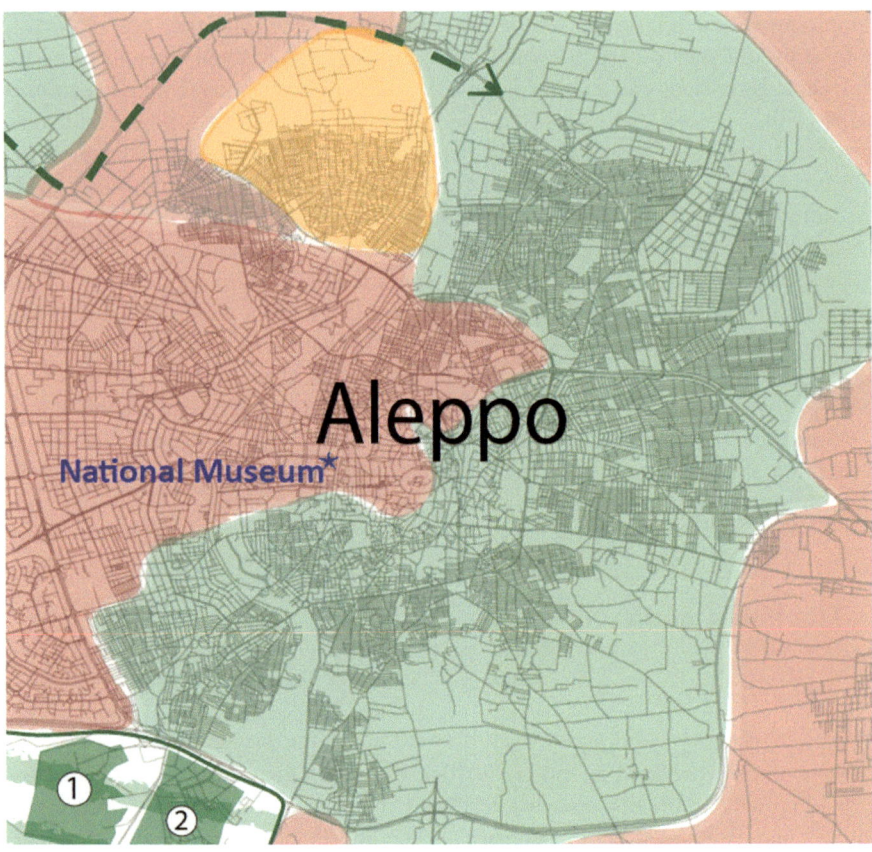

Fig. 1 Urban area of Aleppo as of 31 August 2016. In green the area controlled by the "rebels" in red the ones controlled by the government. Basic map taken from (The Aleppo Project Rebels Break the Siege 2023), in blue the location of the building of the National Museum, in the government controlled area, but very close to the frontline

and they provide a completely new social face of the city itself. In addition, the major religious/ethnic groups that made Aleppo before the conflict, such as Sunnis, Circassians, Curds, Alawis, Christians and Armenians, are now still represented in the city but with different proportions than those of 2011 (Fig. 2).

Since 2017 several NGOs and foundations both national (Syrian Trust for Development) and international (Aga Khan Trust for Culture, UNDP) as well as local institutions (DGAM) planned and, in part, carried out several restoration and rehabilitation projects on the city's cultural heritage (Directorate General for Antiquities and Museums 2019) targeting the monuments that were considered the most relevant ones according to the statutes and aims of each institution. The reopening of the National Museum in Aleppo in 2019 was one among these.

Fig. 2 2019 view of the buildings outside the entrance to the citadel at school drop-off time. ©
Aleppo project

As the war progressed, Syria entered an economic crisis which worsened in 2019,
primarily due to the Lebanon crisis and subsequently to the Cesar Act. Lebanon
was considered the main intermediary in relations between Syria and the external
world starting from the 1950s, especially in reference to the sector banking, as it
was the place where Syrian businessmen and private individuals deposited their
savings and purchased dollars (Yazigi 2020). However, in 2019 Lebanon was forced
to face an economic and financial crisis: banks started to limit the sale of dollars,
the price of US currency rose on the Lebanese foreign exchange market and the
accounts in the Lebanese banks were blocked. This crisis affected imports of primary
(wheat) and industrial (raw materials) goods. In 2020, the Cesar Act went into
effect imposing coercive measures on trade and imports. The expected post conflict
economic recovery never happened reducing the operability in cultural heritage
projects to few international actors and mainly local national institutions sometimes
in competition with each other.

4.2   The Reopening of the Museum

Thanks to UNDP in conjunction with the DGAM and with the financial support
of Japan, in 2017–19 a large part of the debris in the museum area was removed,
the entrance was restored along with the Prehistoric and the Mari Halls (two out

194 M. Pucci

Fig. 3 November 2019, official reopening of the museum with international guests © Aleppo project

of the eight exhibition halls), and half of the roof insulated, more than 2/3 of the building was and is still left without supplies. Most importantly, no restoration has been carried out on any of the sculptures damaged by the war; they are still scattered in the partially abandoned halls (Fig. 3).

Museography remained unchanged from the original 1961 plan, few artefacts were brought back from the storage in Damascus, the flooding problem in the underground storage facility remained unsolved, the other six exhibition halls were left in a post-conflict state exposing the monumental sculptures as well as thousands of artefacts rescued from the flooded storage and kept in there to the action of weather and animals. Most parts of these monumental structure were and still are a landmark of the museum, representing the core of the collection and demonstrating the rich past specifically of the Aleppo region. Sites located in northeast (Tell Brak) or in southeast Syria (Mari) have been chosen among the most representative of the national history in the newly restored exhibition halls; they however show the cultural heritage of regions, where the governmental control is very weak or almost absent and with no the connection to the city and its cultural heritage. However, the reopening fulfilled a symbolic purpose that includes both politics and society: reopening the second museum of Syria in a city that underwent such extended damages on the one hand surely provided a sort of "new start"; keeping the museography intact gives a sense of normalization of the functions of the city; opening the halls to the local community and the few tourists contribute to re-establish a connection between the local population and the national past. Schools begun to visit the museum again, local curators went from being heritage guardians to heritage communicators.

5 The Activities of the University of Firenze at the National Museum of Aleppo

5.1 2021–2022 Still Standing Project

After the reopening of the National Museum in 2019, it became evident that the archaeological heritage preserved in it still needed attention. As mentioned above, the monumental sculptures were neither restored nor put on display but simply left in the dismantled halls. Some of these were found in excavations carried out from the 1920s and are the product of the local traditional craftsmanship of stone carvers. All of them show damages from gunfire, blast, decay and lack of maintenance. This state of "disruption" of the national museum closely mirrors the current conditions of Aleppo's society, which maintains its various and multifaceted elements, but strives to reestablish internal bonds. Besides those monuments that are directly connected to the Aleppo region, the museum hosts monuments, and artefacts from regions where many refugees witnessed the progressive destruction of their archaeological heritage (as an example the 'Ain Dara temple in the Afrin valley, bombed in 2017). The heritage as well as the people coming from these areas show a resilience through conflict and crisis, they are still standing (Fig. 4).

The idea that restoring and reinstalling the archaeological materials specific to this region and neighboring areas, may contribute, at least partially, to the restoration of the identity of the Aleppo community was the core concept of the pilot project "Still standing" carried out in 2021 by the University of Firenze in collaboration with the DGAM; it focused on the restoration of one monumental statue (ninth century BCE) damaged by bomb shelling and found seventy years ago in the urban area of Aleppo. On October 4th we celebrated together with the Museum staff, representatives from the DGAM in Damascus, representatives of several groups of the local community the new-exhibit of the "king" at the entrance hall of the building (Pucci 2022, 2023).

This experience, and the subsequent damage assessment carried out in 2022 with the collaboration of ReStruere (Firenze), let us to better understand the problems and needs of the building and its collections, the availability of materials and know-how in Aleppo and the potential economic development that a complete restoration of the Museum could bring to the citizens. On the other side, meeting, interviews and debates with local curators, visitors, professionals working on different aspects of safeguard and conservation, materials analysis, tourism, and education clearly showed a basic need: to work on a museography that would not "re-create" the same concept of the National Museum as it was conceived before the war, but rather to

Fig. 4 October 2021. Re-exhibit of the 'Ain et-Tell Statue with representatives of the local community and authorities. guests © Aleppo project

embed in the museum narrative the dramatic events that affected the city in the past 15 years and the ever changing social complexity that the city is now representing.

5.2   The 2023 Earthquake

The rehabilitation project of the building and its collection faced at the beginning of this year a new challenge: the consequences of the earthquake that occurred on February to 6th, 2023. These consequences were twofold: (1) Visible damages to the infrastructure and completely absence of electricity and ventilation. (2) The museum curators and employees in part left their houses immediately after the event fearing further earthquake events and have been using the museum structure as a provisional stay for four months after the earthquake, ascribing to the building the value of a "safe space".

As a result of these events, in 2023 the project focused on earthquake relief measures on the cultural heritage (physical damages to the structure and to the collection) and aimed also at providing direct relief to the local community. In the earthquake damage assessment carried out on the structure in June 2023 together with the DGAM and ReStruere (Fig. 5), it has been possible to state that the building suffered some structural damages that did not affect its stability, but consistently

Fig. 5 June 2023. Post-earthquake damage assessment on the first floor of the museum building with ReStruere and Unifi team. © Aleppo project

weakened the building that would not be able to face a second, even lighter earth-quake event. Moreover, analyzing in detail the structure it became evident that if we would intervene to make the building an earthquake-proof structure, the museum could become also a place that local communities may use as shelter in the event of an earthquake. This element would add much value to the role of the museum inside the community. Therefore, after some emergency interventions to close the shattered windows and remove the rubble, the university of Florence begun with the DGAM to sample the building materials to plan structural intervention that would keep the structure stable in case of an earthquake event. The museum project would become a "cultural construction site" setting the museum not only as a cultural heritage but also as an institution actively contributing to earthquake relief. Thanks to the financial support of Al-Aliph foundation and the cooperation of Terres des Hommes—Italy first earthquake relief intervention begun in October 2023 and will continue in 2024, guaranteeing power supply for the whole structure (including the pumps) with solar energy, repairing the earthquake damages and planning structural interventions to make it earthquake proof.

5.3 The Museography Concept

"A museum is a not-for-profit, permanent institution in the service of society that researches, collects, conserves, interprets and exhibits tangible and intangible heritage. Open to the public, accessible and inclusive, museums foster diversity and sustainability. They operate and communicate ethically, professionally and with the participation of communities, offering varied experiences for education, enjoyment, reflection and knowledge sharing" (ICOM 2023).

The museography concept of the new National Museum of Aleppo should consider the long history of its territory and the population living in it. Its main aim is to provide the city of Aleppo, its citizens as well as Syria and the Syrians with a modern, inclusive, and sustainable national museum for archaeology, based on a bottom-up approach to experience the archaeological collection and use it as a medium for education and peace building. Each bottom-up approach requires a constant dialog in the making, that means that a draft museography idea will be discussed, changed and redrafted on the making, and we are still at the first steps of comparison, discussion and negotiation not only with the museum staff, but mainly with all local actors (stakeholders, participants, facilitators, volunteers, students) who may contribute to shape the narrative (Fig. 6).

In the first draft, modernity is mainly represented by a combination between a "traditional" educational approach with a thematic organization of the exhibit rooms. Different themes, that are related to human life, such as domestic activities, ritual spaces or funerary customs, will be presented, in chronological order, using artefacts and monuments that will show to the museum guests not only change and continuity of traditions over time, but also the ability of Syrian territory and of Aleppo in particular to offer an inclusive space, where different communities could live together keeping their traditions and beliefs and sharing everyday customs and activities in the past as well as today in the multifaceted Syrian territory. Using the same structure of the museum through a profound reshaping of the spaces and using local materials and expertise is considered the best approach to guarantee both durability and sustainability of the project. Moreover, the museum structure and its location has

Fig. 6 BIM model of the museum structure © ReStruere

become over the years a marker in the urban topography of the city, keeping its original location but rearranging its collection signals the local community innovation in continuity. Renovation of the structure, restoration of the collection and rearrangement of the materials following both an educational and thematic approaches are the activities that should transform the museum into a shared safe space, a place for training, discussion, confrontation, i.e. a cultural building site.

References

Abdulkarim M, Cunliffe E (2022) The Syrian example. In: Cunliffe E, Fox P (ed) Safeguarding cultural property and the 1954 Hague Convention: all possible steps. Boydell & Brewer, Woodbridge, pp 181–204

Abt J (2006) The origins of the public museum. In: Macdonald S (ed) A companion to museum studies. Blackwell Publishing Ltd., Oxford, pp 115–134

Bandarin F (2022) The destruction of Aleppo: the impact of the Syrian war on a world heritage city. In: Cuno J, Weiss H (eds) Cultural heritage and mass atrocities. Getty Publications, Los Angeles, pp 186–201

Bernhardsson TM (2005) Reclaiming a plundered past. Archaeology and nation building in modern Iraq. University of Texas, Austin

Borghese B (2015) Spotlights on Syria. News Conserv 46:9–13

CEIC Syria Visitor Arrivals. https://www.ceicdata.com/en/indicator/syria/visitor-arrivals. Accessed 22 Nov 2023

Chamonard J (1920) A propos du service des antiquités de Syrie. Syria 1:81–98

Cluzan S, Delpont E, Mouliérac J (1993) Institut du monde Arabe: Syrie, mémoire et civilisation. Flammarion, Paris

Directorate General for Antiquities and Museums (2019) The intervention plan for Aleppo ancient city (UNESCO document WHC/19/43.COM/7A), State Party Report on the State of Conservation (STATE REPORT 7a with annexes). 43rd Session of the World Heritage Committee. UNESCO Baku, Azerbaijan

Dussaud R (1925) Nouvelles archéologiques. Syria 6:291–300

Dussaud R (1931) Nouvelles archéologiques. Syria 12(2):190–192

Fakhro M (2020) Strategies for reconstructing and restructuring of museums in post-war places (National Museum of Aleppo as a model). PhD thesis, University of Bern

Fakhro M (2021) Protection measures taken by the museum during the Syrian was: the National Museum of Aleppo as a model. In: Jackson H, Jamieson A, Robinson A, Russell S (eds) Heritage in conflict: proceedings of two meetings: "Heritage in Conflict: A Review of the Situation in Syria and Iraq", Workshop held at the 63rd Rencontre Assyriologique Internationale, Marburg, Germany, 24–25 July 2017, and "Syria: Ancient History—Modern Conflict" Symposi. Peeters Publishers, Leuven, pp 43–58

Fortin M (1999) Syria, Land of civilizations. Musée de la Civilization de Québec, Montreal

Gibson M (2008) The looting of the Iraq museum in context. In: Emberling G, Hanson K (eds) Catastrophe! The looting and destruction of Iraq's past. The University of Chicago Press, Chicago, pp 13–19

Gillot L (2010) Towards a socio-political history of archaeology in the middle east: the development of archaeological practice and its impacts on local communities in Syria. Bull Hist Archaeol 20(1):4–16

Hodder I (2010) Cultural heritage rights: from ownership and descent to justice and well-being. Anthropol Q 83(4):861–882

ICOM (2023) ICOM museum definition. https://icom.museum/en/resources/standards-guidelines/museum-definition/. Accessed 05 Nov 2023

Kalshoven F (2005) The protection of cultural property in the event of armed conflict within the framework of international humanitarian law. Mus Int 57(4)

Maqdissi MA (ed) (2008) Pionniers et protagonistes de l'archéologie syrienne 1860–1960, d'Ernest Renan à Sélim Abdulhak. Documents d'Archéologie Syrienne XIV. Direction générale des antiquités et des musées, Damas

Munawar NA, Symonds J (2022) Post-conflict reconstruction, forced migration and community engagement: the case of Aleppo, Syria. Int J Herit Stud 28:1017–1035

Museum DN (ed) (1996) Exposition Syro-Européenne d'Archéologie: miroir d'un partenariat. National Museum of Damascus, Damascus

Normand R (1921) La Création du Musée d'Adana. Syria. Archéologie, Art et histoire, pp 195–202

Ouahes I (2018) Syria and Lebanon under the French mandate. Cultural imperialism and the workings of empire. I.B.Tauris, London, New York

Ploix de Rotrou G (1932) Le Musée National d'Alep. Catalogue Sommaire. Revue Archéologique Syrienne II, pp 35–83

Pottier E, Dussaud R, Migeon G (1920) Avertissement au lecteur. Syria 1(1):1–2

Pucci M (2022) Another Ruler in Bit Agusi? The 'Ain et-Tell Statue at the National Museum of Aleppo. Studia Eblaitica 8

Pucci M (2023) Aleppo e il suo museo. Tra guerra e terremoto. Archeologia viva 220

Qassar H (2015) Fostering values to protect cultural heritage in Syria. News Conserv 46:24–25 (2015)

Rossi M (2011) The museum display: theoretical implications. In: Rossi M (ed) Archaeology of Cooperation Afis—Deinit and the Museum of Idlib. Activities in the frame of the MEDA project. Tilapia, Napoli, pp 297–329

Stein GJ (2015) The war-ravaged cultural heritage of Afghanistan: an overview of projects of assessment, mitigation, and preservation. Near East Archaeol 78(3):187–195

The Aleppo Project Rebels break the siege. https://www.thealeppoproject.com/august2016/. Accessed 22 Nov 2023

UNESCO (2023) UNESCO reports on extensive damage in first emergency assessment mission to Aleppo. https://www.informea.org/en/unesco-reports-extensive-damage-first-emergency-assessment-mission-aleppo. Accessed 22 Nov 2023

Varutti M (2014) Museums in China: the politics of representation after Mao. Boydell & Brewer, Woodbridge

Virolleaud C (1924) Rapport sur les recherches du service des Antiquités de Syrie en 1924. Comptes rendus des séances de l'Académie des Inscriptions et Belles-Lettres 68(4):280–281

Wartke RB (2005) Sam'al: Ein aramäischer Stadtstaat des 10. bis 8. Jahrhunderts v.Chr. und die Geschichte seiner Erforschung. Zabern, Mainz

World Bank Group (ed) (2017) Syria damage Assessment of selected cities: Aleppo, Hama, Idlib International Bank for Reconstruction and Development/The World Bank Washington

Yazigi J (2020) Syria's Growing Economic Woes: Lebanon's Crisis, the Caesar Act and now the Coronavirus. Arab Reform Initiative

International Cooperation in the Field
of Cultural Heritage

The Italian Agency for Development Cooperation. Culture as a People-Centered Path to Development: Best Practices from Fragile Areas—An Overview of Actions in Jordan, Palestine, Lebanon, Afghanistan, Pakistan

Paola Abenante and Fabio Strinati

Abstract This text outlines the Italian Agency for Development Cooperation's approach to culture as a tool of *Prosperity*, which contributes to the creation of favourable conditions for socio-economic development. It argues that such an approach must necessarily be people-centred and attentive to ownership and inclusivity, especially when dealing with culture, understood as a process and product of human expression. To promote culture in development involves prioritizing people's meanings, goals, tools and, more broadly, societies' right to self-determination and to establishing their own development priorities. An attention to culture and cultural diversity in development favors a process in which multiple parties come together and leverage their respective tools to achieve shared and more sustainable results. An array of projects carried out by AICS in the Region focused on in this volume, exemplify how, from a methodological and operational standpoint, Italian actions are designed to generate economic activity and participation *for* and *through* culture, placing people at the center.

Keywords Culture · Socio-economic development · People-centered approach · Italian cooperation

P. Abenante (✉) · F. Strinati
Italian Agency for Development Cooperation, Via Cantalupo in Sabina 29, 00191 Rome, Italy
e-mail: paola.abenante@aics.gov.it

F. Strinati
e-mail: fabio.strinati@aics.gov.it

© The Author(s) 2024
M. Loda and P. Abenante (eds.), *Cultural Heritage and Development in Fragile Contexts*, Research for Development, https://doi.org/10.1007/978-3-031-54816-1_13

1   AICS Approach to Culture and Development

Culture is a priority for AICS and the Italian cooperation. Since its inception, the Italian Agency for Development Cooperation has been investing in the cultural sector, disbursing approximately 90 million euros in grants over the first seven years of its activity (2016–2022). These resources support projects carried out in four main areas of intervention, which include the protection and safeguarding of tangible and intangible cultural heritage, the development of cultural and creative industries, the development of sustainable tourism, and the promotion of education and participation in the cultural sphere.

More specifically, Italian cooperation values culture as a tool of *Prosperity*, one of the four pillars of the Agenda 2030, focused on "improving livelihoods and ensuring that all can enjoy prosperous and fulfilling lives" (https://sdgs.un.org/2030agenda). On the path towards achieving prosperity, the economic and social dimensions of development are intertwined and not easily distinguishable, and this is all the more evident when focusing on culture as a process and product of human expression. Indeed, expanding on the UNESCO 2003 definition of intangible cultural heritage,[1] culture may be understood as the ongoing process through which people interpret the reality they live in, as well as the expression and representation of such reality.

Cultural goods and services are the by-products of this process, since they are directly linked to people's creativity, skills and knowledge. This renders cultural products economically sustainable and likely to favor the diversification of the market, as they are resistant to automation, due to the intellectual labor inherent in the cultural and creative process of production.

On the other hand, from the standpoint of social sustainability, supporting culture and the multiplicity of cultural expressions means supporting the identities and values of communities and of society at large, in a way that strengthens social cohesion and resilience. Moreover, such an understanding of culture necessitates recognition of the diversity of cultural expressions in relation to different contexts and communities, and peoples' diverse positioning in relation to what is expected from development.

By combining creative skills, production and distribution activities of cultural goods and services, the cultural sector generates employment and income, and contributes to building a democratic public sphere, open to the expression of creativity and cultural diversity. Therefore, to promote culture means to promote development as a people-centered process that gives priority to people's meanings, goals and tools and, more broadly, prioritizes societies' right to self-determination and to establishing their own development priorities, from the perspective of ownership and of inclusive growth.

An attention to culture and cultural diversity in development favors a process in which multiple partners come together and leverage their respective tools to achieve

---

[1] In the UNESCO *Convention for the Safeguarding of Intangible cultural Heritage* (UNESCO 2003), ICH is defined as a set of abilities, knowledge and values through which communities and more at large societies, from generation to generation, know and interpret the world they live in, as well as the objects, artefacts and spaces associated therewith.

shared results. Beyond being a means of prosperity, culture plays an important role in accomplishing the change of register that underlies the 2030 Agenda, encouraging the transformation of aid relationships between donor and beneficiary into cooperation practices among partners. Culture-based development is the bedrock of cooperation as a process of mutual learning and knowledge exchange, a form of partnership in which partners grow together, with each party contributing according to its own means.

2 Socio-economic Development: Ownership and Inclusivity in the Field

The path to achieving sustainable social and economic development goals is a complex one, and respect for ownership and inclusivity is an ongoing challenge, which strives to recognize and leverage the diverse and ever-changing interests, needs and complementary roles of the multiplicity of stakeholders involved. Such a challenge is all the more evident when working with culture, which is a dynamic process of defining and re-defining meaning, carried out by communities. From a methodological and operational standpoint, and with regard to ensuring a people-centered approach that aims at ownership and inclusivity, the main features of AICS' approach in designing and implementing actions on culture are ground-based programming, rooted in an extended presence of regional offices in the field, and the implementation of partnerships with local public and private stakeholders. Such an approach ensures a thorough knowledge of the territory, of its stakeholders, of its tangible and intangible cultural heritage and its needs. Moreover, the presence of AICS experts and field officers on the ground ensures the ongoing monitoring of the projects, not least in view of their necessarily continuous adaptation and re-alignment to the changing contextual conditions of partner societies and their needs.

Among the various areas of intervention, the protection of cultural heritage is a strength of Italian Cooperation, due both to the excellence of the Italian system of expert partners in the field of conservation, restoration and valorization of cultural heritage, and to Italian Cooperation's long-lasting experience in projects concerning cultural heritage. The Italian approach boasts a history that digs its roots in the *Culture Counts* conference held in Florence (1999), co-promoted by the World Bank and the Italian Government in which, in an innovative way at the time, the role of culture in sustainable development was affirmed as a multifaceted resource.

Following this approach, AICS' actions on culture aim at increasing, on the one hand, cultural employment, cultural businesses and public and private financing to the culture sector, on the other hand they aim at supporting cultural knowledge, cultural participation and participatory processes. Finally, from a wider perspective, the Italian cooperation supports the economy of culture by developing and reinforcing the governance of culture. Indeed all of the projects presented in the following of this contribution foresee the elaboration, or the support to the elaboration, of local development plans or sector management plans by and in collaboration with local authorities.

3 Generating Employment *For* and *Through* Culture

AICS' projects in the field of Culture and Development directly and indirectly generate job opportunities. Such opportunities are either for cultural professionals in the heritage and tourism sector, as well as within other connected sectors, or for non-cultural professionals and unskilled laborers in the heritage and tourism sector.

Most of AICS' projects leverage existing skills in the culture sector through capacity building and professional training, mainly targeting specialized technical personnel in public national and local institutions, such as local ministries of culture, universities and specialized centers. Recurring training topics relating to the thematic sector of the Safeguarding of Cultural Heritage are: restoration, museum setup, heritage digitization, valorization tools and practices, as well as training in business management to support the creation of cultural enterprises.

Best practices in capacity building and professional training, aimed at expanding the cultural job market within the Region focused on in this volume, are the interventions AICS carries out in Lebanon and in Jordan. These interventions support the creation of institutions, national and regional in their scope, which specialize in the restoration and recovery of cultural heritage and in determining the curricula of local professionals on the field. Such projects, centered on training activities for professionals, have a snowball effect, accelerating the expansion of the cultural job market. AICS Amman's project *Creation of a Regional Institute for the conservation and restoration of cultural Heritage in Jerash* is an example. Carried out by AICS Amman, in partnership with the United Nations Office for Project Services (UNOPS) and the University of Roma Tre, and in close coordination with the Jordanian Department of Antiquities, this project supports and assists Jordanian institutions which are active in the conservation and valorization of archaeological sites, historical monuments and museums. The Regional Institute of Conservation and Restoration of Jerash is set to be a specialized center. As such it will be fully capable of ensuring the development of new skills and greater professional capabilities in the Jordanian system of management, conservation and restoration of cultural heritage, both at the national and regional level. The Institute will train public and private professionals coming from all over the Middle East, widening the cultural job market.

The *Program in support of the socio-economic recovery in the protected area of Shobak castle* is described in the University of Florence's contribution to this volume. It represents another best practice in the support of local and national public sector authorities, implemented in Jordan. The Shobak program supports the socio-economic recovery of local communities through the rehabilitation of the archeological site, the touristic trail and small units around the castle. As such, it aims to generate potential livelihood opportunities for local communities. Within the larger framework of this program, the University of Florence carries out—beyond technical and conservation actions as detailed in the article—an intense training and capacity-building program, covering a vast range of disciplines linked to conservation, and directed at a number of professionals—archaeologists and engineers—associated with the local Department of Antiquities. The training aims at upskilling public professional experts who will be in charge of the management and conservation of the area after the conclusion of the project, thus ensuring its sustainability.

The CHUD program (*Cultural Heritage and Urban Development*) implemented in Lebanon, represents another best practice in the assistance and support to national partner institutions. The program seeks to preserve and restore the country's cultural heritage by intervening on the historic cities of Baalbek, Byblos, Saida, Tripoli, and Tyre, as described in AICS Beirut contribution. The CHUD program is not only an important example of support to the public sector, but is also an example of a program that fully respects the country ownership principle. Indeed it is a soft loan program that originates from a decision made by the Government of Lebanon to invest in heritage preservation and urban development as sustainable development sectors. Such a decision was financially supported by the World Bank and two bilateral donors, France and Italy. The Italian component contracted UNESCO to provide advice and technical assistance to the Lebanese Ministry of Culture, regarding the two World Heritage sites of Tyre and Baalbek. This advice and assistance covered the monitoring and supervision of works in both sites, while helping to meet the necessary World Heritage requirements.

With regard to the involvement of civil society, AICS actions work to generate direct job opportunities as well as spin-off employment opportunities in the service economy linked to culture, i.e. the tourist industry, while building or supporting local cultural industries. Among the operational means to achieve direct job opportunities, particularly in the fragile areas dealt with in this volume, *cash -for- work* allows for the mobilization of unskilled workers. It also indirectly contributes to raising awareness and encouraging inclusivity and participation in the field of culture. Both the contributions to this volume by AICS Beirut and AICS Amman describe how cash-for-work has been the bedrock of interventions in the *Cultural Heritage and Urban Development* program (CHUD) as well as in the *Petra Siq Stability Program.* The Petra program, beyond the structural stabilization of Petra's World Heritage Site, carried out in the first three phases of the project, has a fourth phase specifically aimed at community engagement in risk prevention. Eighty-five skilled and unskilled women and men, mainly young people, were employed through cash-for-work on-site, both in risk mitigation and in the cleaning of the Siq slopes. These people were identified through a socio-economic vulnerability assessment carried out in the nearby villages. The project provided financial support, contributing to the stability of households and communities facing particularly harsh socioeconomic conditions in the immediate post-COVID period. Beyond the immediate financial support, workers received technical training and gained experience, thus upskilling for future employment. A further positive outcome of the local workers' active engagement has been raising community awareness about the social and economic value of cultural heritage (Fig. 1).

Fig. 1 Petra Siq Stability Program—Jordan

For another example of the creation of both direct and indirect opportunities for civil society within the region, one can turn to Palestine. AICS has a project underway, implemented by UNESCO, focusing on the World Heritage Site 'Palestine: Land of Olives and Vines—Cultural Landscape of Southern Jerusalem, Battir', located a little south-west of Jerusalem and to the West of Bethlehem. This area boasts a major Palestinian agro-cultural landscape, across which agricultural traditions are still practiced, and constructions such as ancient dry wall terraces are still being utilized, while the natural water supply is harnessed through a complex and unique irrigation system.

Poor market access, high transaction costs and reduced freshwater availability, as well as a decline in sociocultural traditions and agricultural practices, has caused a socioeconomic crisis characterized by a general decline in agricultural livelihoods: agricultural practices are progressively abandoned, leading to increasingly degraded land and terraces. The project aims at safeguarding Battir's unique agricultural landscape in two ways: supporting the most severely affected farming families in the area, by revitalizing traditional agricultural techniques and promoting local products, while at the same time supporting the larger community by promoting sustainable tourism. The involvement of civil society organisations (CSOs) is a priority in the development of the project, which directly addresses farming families in Battir and Hussan villages, active male and female agricultural cooperatives, in addition to cultural tourism operators/service providers in Battir. In fact, the direct beneficiaries include farming families who rely entirely or partially on agricultural production activities to make a living, and have the potential to effectively engage in production and marketing through the cooperatives.[2]

4 Generating Participation *For* and *Through* Culture

From the perspective of social development, AICS promotes projects that are embedded in society, by encouraging participation on multiple levels. The participation of local and national authorities, as previous examples have highlighted, is ensured by the establishment of partnerships at the onset of every action. The CHUD program in Lebanon boasts a very broad partnership that ensures national ownership, by involving the Lebanese Government through its dedicated bodies. Among these are the Council for Development and Reconstruction, the Ministry of Culture and the Direction for Antiquities, the Ministry of Public Works and the Direction for Urbanism, the Ministry of Tourism, as well as local public bodies such as the different municipalities of the towns participating in the project. Beyond the public sector the CHUD program involves local CSOs, universities and research centers, on the civil society level.

At the same time, by listening to the different voices that are necessarily involved in actions concerning cultural heritage, AICS' projects encourage participatory processes within civil society, starting from careful assessments of local socio-cultural dynamics on the field. In this sense, best practices are carried out by the University of Florence in Bamiyan and Herat, and described in detail throughout this volume. The contributions analyze the socio-cultural context by highlighting the main dynamics that shape people's understandings, representations and attitudes towards the cultural heritage they live amidst. Local practices such as land titling, gender or ethnic dynamics influence how local communities relate to the conservation and management requirements of a World Heritage Site. Such understanding on behalf of project-implementing bodies is essential in drawing up heritage protection solutions that are respectful of local communities' attitudes, therefore effectively participatory, inclusive, and sustainable.

Essential to a participatory approach, and tightly linked to the above, are actions of raising awareness that address local communities. According to UNESCO, AICS' project *Radio Education* in Pakistan is to be considered a best practice in the Region. The main objective of this initiative is to improve access to education for children and young people, their families and their respective communities in the most marginalized and isolated provinces of Pakistan. This is done through radio programs that convey educational messages. The aim is to promote the spreading of knowledge while raising awareness on various issues among which are included cultural heritage and environment protection, disaster risk reduction and global citizenship. Selected schools and communities received radios built specifically for the project, thanks to contributions offered by private companies, all of which have been equipped with

---

[2] Source UNESCO Project Document for AICS.

Fig. 2 Radio education—Pakistan; project implemented by UNESCO

rechargeable solar-powered batteries to enable functionality even in areas out of reach
of electricity. The radio programs have been broadcasted daily and have allowed for
the active participation of primary school children and teachers (Fig. 2).

The beneficiaries of this project have been actively involved throughout the development phase (when the content of the radio programs was decided) and in the broadcasting phase when they had the opportunity of interacting and participating live. Of
those who benefited from the project, there were around 30,000 primary school
children (from 5 to 14 years of age), approximately 400 schools, with the direct
involvement of about 500 teachers. Indirect beneficiaries included families and their
communities, specifically from the areas of Khyber Pakhtunkhwa, Sindh, Punjab,
Balochistan, Gilgit-Baltistan, Islamabad Capital Territory and Pakistan Adminis-
tered Kashmir, but also the departmental administrations at the provincial level, and
the federal government, which deals with education, culture and information.

5   Conclusions

The projects briefly described above are designed to have an impact, in the long term,
by promoting culture as an enabler of prosperity, and by contributing to the creation
of favorable conditions for socio-economic development. The promotion of human
capital and expression carried out by cultural actions can indeed positively affect
governance, democracy and stability.

In the countries where it operates, AICS supports national and local institutions in strengthening technical and professional skills and in building inclusive management models. AICS also supports non-profit and for-profit organizations, small and medium-sized industries, as well as individuals, by investing in innovation, in knowledge and in the valorization of civil society, with the aim of capitalizing and leveraging the plurality of society's skills and voices.

Supporting the governance of culture by assisting the development of cultural policies, management plans and the upskilling of local institutions and professionals, builds upon such an approach. Cultural heritage management plans, grounded in technical and scientific research, aim at balancing out the different positions of stakeholders in the field, so as to find a common ground with regard to heritage protection and safeguarding, in an attempt to enhance the universal value of culture.

As recognized in the OECD Peer Review (OECD, 2020), Italy has the ability to add greatly to this field, given the excellence and expertise of the Italian system in cultural heritage.

AICS' actions are implemented by bodies—universities and international organizations—that have a profound knowledge of the territory in which they operate, also thanks to their long-term presence on the field. AICS contributions as follow, in line with the approach of this entire volume, wish to capitalize on such deep knowledge and broad experience, in order to reflect on and explore the long-term impacts that research and cultural heritage protection activities can lead to, focusing on the relationship with civil society, and on the issues of participation and ownership.

The goal is to understand how efforts carried out in the cultural sector can contribute not only to the preservation of material heritage, but also to the enhancement of human capital. From this perspective, the following contributions provide examples of actions in the field, highlighting the main tools and approaches in interacting with national bodies and local communities, to improve the resources, support skills and knowledge of the people who live and identify with the heritage.

References

OECD (2020) Peer Review dell'OCSE sulla cooperazione allo sviluppo: Italia 2019. OECD Publishing, Paris. https://doi.org/10.1787/e752c41e-it

UNESCO (2003) Convention for the safeguarding of the intangible cultural heritage. https://ich.unesco.org/en/convention. Accessed 22 Nov 2023

United Nations. Transforming our world: the 2030 agenda for sustainable development. https://sdgs.un.org/2030agenda

The Italian Agency for Development Cooperation in Lebanon

Maria Luigia Calia and Alessandra Piermattei

Abstract The actions presented by AICS Beirut add to the largely positive impact that the Italian Cooperation has had in safeguarding the cultural and natural heritage in the fragile context of Lebanon. These Italian actions are implemented within the *Cultural Heritage and Urban Development Program* (CHUD), launched in Lebanon as a driver for development and for bolstering national and international tourism. The Cooperation's activities contribute to the mitigation of heritage degradation risks, while at the same time supporting the upskilling of a large number of local professionals, trained to intervene effectively in securing and preserving Lebanese cultural heritage. In committing to the preservation and valorisation of Lebanese cultural and natural heritage, there is no escaping the structural problems of the monuments, cleaning parts that need curative activities, conserving architectural and decorative elements, consolidating the structures, and making the sites accessible to the local population and visitors. The actions are tailored to the specific goal of enhancing cultural tourism, which is a major source of income in Lebanon, and provides many job opportunities, revitalizing local economies. At the same time, a significant effort has been made to forge a connection between community and territory, in order to increase a sense of belonging, by planning both information days and workshops.

Keywords Heritage preservation · Culture driver for development · Sustainable inclusive tourism

M. L. Calia
Senior CHUD Program Officer, Beirut, Lebanon

A. Piermattei (✉)
Head of office - AICS Beirut, Beirut, Lebanon
e-mail: alessandra.piermattei@aics.gov.it

© The Author(s) 2024
M. Loda and P. Abenante (eds.), *Cultural Heritage and Development in Fragile Contexts*, Research for Development, https://doi.org/10.1007/978-3-031-54816-1_14

215

1 Premise

The exceptional cultural and living heritage in Lebanon creates incredible opportunities for sustainable development and tourism. Tourism is a major source of income in Lebanon given the availability of tourism-related work and, naturally, wide-scale employment in this sector helps revitalize local economies, bringing money into the country. Since the financial crisis in 2019, the Lebanese economy has been in a state of fragility, the crisis having had an impact on much of the country; on its heritage, its economic activities and on its tourist industry. Given this situation, providing support to cultural heritage and tourism in Lebanon is all the more relevant.

Italian support for Lebanese cultural heritage, through AICS, aims at salvaging sites at risk. To do so, the Italians have contributed with approximately 13,828,000 euros to this end. The Italian Cooperation has intervened on various Lebanese sites of historical, cultural and archaeological interest, including three World Heritage Sites (Baalbek, Tyre, Qadisha Valley). Other sites of intervention, neglected due to the country's modest financial resources, are the Saida Land Castle in Sidon, the Khan el Echle Caravanserai in Saida, the Chamaa Castle, the historic Town Hall in Baalbek, the basement of the National Museum in Beirut, and the Sursock Museum in Beirut.

The main activities for preserving and promoting cultural heritage are outlined in the framework of the *Cultural Heritage and Urban Development Program* (CHUD), to which Italy has contributed with a soft loan of 10,228,000 euros and a grant of 2,200,000 euros. Another important intervention is the project for the restoration and valorisation of Wadi Qadisha, entrusted to UNESCO with a contribution of 500,000. Wadi Qadisha is the most important early Christian monastic settlement and is included in the UNESCO World Heritage List. The Italian Cooperation is also involved in the restoration of the Sursock Museum in Beirut, contributing with a grant of 1,000,000 euros. This intervention funds the restructuring of the museum, its electro-mechanical infrastructure, doors and lighting, in order that it might be reopened after the damages caused by the Beirut Blast. The Italian Cooperation is also funding the project *Cultural Religious Tourism*, thanks to a grant of 400,000 euros. The aim of this intervention is to carry out a mapping of the rich and diverse Lebanese sites of cultural, historic, religious, and archaeological value and to promote them as a prime global, cultural and religious tourist and pilgrimage destination.

2 The National Museum Restoration

The restoration and restructuring of the basement of the National Museum in Beirut, working off a grant of 1,000,000 euros, made possible the public reopening of this part of the museum, closed since the beginning of the Lebanese civil war in 1975. The project was able to renovate the basement completely, as well as to set up a modern display of its funeral art collections. The Italian commitment to support the

National Museum of Beirut, technically and financially, has led to the restoration of a total area of 550 square meters, upon which a new and modern exhibition was set up in the basement, designed by an Italian architect, to celebrate the history and culture of the country. The project, managed directly by AICS Beirut, was carried out alongside a group of experts, and succeeded in mobilizing a number of specialists, capable of restoring and preserving the collections to be displayed. The project provided technical assistance to the Lebanese General Direction of Antiquities for the restoration and conservation of monuments such as the second century AD Tyre Tomb. Frescoes were restored and relocated to a specific space with the help of ICCROM experts. The restoration and conservation of the mummies, dating back to the thirteenth century, and previously contained in the Assi el-Hadath cave in the Qadisha Valley, was carried out with the help of experts from the specialised EURAC Research Institute in Bolzano. The museum boasts an extraordinary collection of anthropomorphic sarcophagi, found in Sidon and dating back to the fifth/sixth century AD. The scenography was put together by an Italian architect. Since the inauguration of the restored basement in 2016, the Museum has welcomed high numbers of visitors, while ticket sales are progressively rising. Italy also supported the publication of the National Museum Guide in four languages (Arabic, English, French and Italian).

3   The CHUD Program

The Government of Lebanon decided to invest in preserving heritage and in urban development as sustainable development sectors and, in 2003, with the financial support of the World Bank and bilateral donors (France and Italy), it launched the CHUD program, aimed at preserving and restoring the country's cultural heritage in the historic cities of Baalbek, Byblos, Saida, Tripoli, and Tyre. The Council for Development and Reconstruction (CDR) contracted UNESCO in order to provide advice and technical assistance to the Lebanese Ministry of Culture for the two World Heritage sites (Tyre and Baalbek). In particular, the contract aims to assist the General Direction of Antiquities in the monitoring and supervision of projects in Tyre and Baalbek, as well as meeting the World Heritage consultation requirements, monitoring the conservation measures to be implemented, providing reports and organising workshops and events on heritage intervention methodologies. The CHUD program puts culture at the heart of city reconstruction and recovery processes in the wake of crisis. The program was designed to provide operational lessons on how to bolster local economic development, enhance citizens' quality of life, and improve the conservation and management of Lebanese cultural heritage.

The intervention started with assessments and surveys on the state of conservation, collecting all necessary documentation and carrying out archaeological investigations. All this information, useful in updating the historical interpretative data, has informed conservation plans concerning surfaces and structures. As a consequence, when the works started, techniques of restoration and consolidation were tailored to the specific needs of the site. Moreover, an important effort has been made to respect international protocols, by applying non-invasive methods so that the integrity of the sites and monuments could be maintained. Alongside this, the intervention was able to increase local know-how in the protection and management of heritage, provide infrastructure and services, train unskilled workers on the job, raise awareness of heritage value, and enhance participatory processes, accessibility and inclusivity for all.

The creation of a large network of partners is an additional achievement of the intervention. The partnership involves: the Lebanese Government—CDR, the Ministry of Culture and General Direction of Antiquities, the Ministry of Public Works and the General Direction of Urbanism, the Ministry of Tourism, the Municipalities, Universities and research centres, citizens of the towns participating in the projects, and civil society organisations.

Moreover, the project involved a number of Italian experts, capitalising on Italian technical and scientific knowledge, as well as best practices in restoring and preserving historical monuments.

In terms of civil participation and of awareness raising, the project also focused on site accessibility and communication strategies. Beyond improving the safety of the sites, new visitor itineraries and walkways have been created with the aim of valorizing the cultural and historical value of the sites and landscape, including in the remote and less known areas. The visitor itineraries have been put together to promote the discovery and restoration of the site and to improve visitor experiences. This can be achieved by presenting the monumentality of the place and placing it in its historical context. Focus has thus been placed on the cultural complexity of the sites, on the multiplicity of historical layers, on the continuity of history, and thus on different interpretation perspectives.

The design of these visiting circuits has been shared and agreed upon among local and national authorities, as well as civil society organizations (Fig. 1).

4 The Baalbek Intervention—Jupiter Temple's Colonnade

The preservation of Jupiter's Temple Colonnade, the most iconic monument within the site of Baalbek,[1] is a best practice in intervention, that capitalizes on Italian methodology and expertise, and serves as a pilot for future interventions on both national and regional cultural heritage sites. The Jupiter temple, within which only six columns remain standing, is twenty meters high, and was once supported by an

---

[1] The Colonnade of Baalbek, the reaming part of the roman Jupiter Temple, is the most visited of all the ancient monuments in Baalbek, not to mention, a Lebanese icon. It has survived earthquakes, reuse, neglect, and abandonment for about two millennia. The magnificent colonnade offers a view of perfection, stylistic composition, and monumentality of the site. For several centuries now, the ancient Roman town of Baalbek, having been excavated in large parts, has presented the visitors with a unique opportunity to discover the grandeur of the constructive techniques, the decorative surfaces, shapes, and details of this grand monument.

Fig. 1 Baalbek—panoramic view of the infrastructure for visitors

entablature of more than five meters. It forms a rectangle of eighty-eight meters by forty-eight, which makes it the largest temple in the Roman world. Its purpose was to show, by inscribing it in stone, the power of the Roman Empire that led Augustus to construct a great sanctuary at Heliopolis; the works began under his reign and continued up until the end of the second century. Celebrating the glory and power of Rome, the temple was built in accordance with the principles that characterize Roman religious architecture and the decorative elements were borrowed from the Greco-Roman ornamental repertoire. However, the organization takes into account the religious usages of Rome.

The project intervened on structural decaying and surface pathologies affecting the Jupiter Temple's Colonnade, applying suitable solutions designed to preserve the integrity of the monument. The conservation was made possible, thanks to the scaffolding that provided the opportunity to conduct a widespread check-up and intervention from up close, and up to the very top of the temple. This phase has been completed and solutions are now being implemented in tackling the most critical problems concerning its stability and preservation (Fig. 2).

The Italian integrated approach, applying the most appropriate techniques and knowledge, and in line with the restorations and conservation charters, was shared and approved of at the international and national level, along with the involvement of ICOMOS/UNESCO experts, who were in charge of checking and preparing independent reports on the project's progress and results. These experts, alongside concerned

Fig. 2 Restoration works, Jupiter Temple—Baalbek

authorities and civil society representatives, have been gathered in order to provide feedback and advice on what was to be done. On the site itself, architects, archaeologists, engineers, conservators, workers, management staff worked alongside one another, dealing with the monument's structural problems, cleaning parts in need of curative action, conserving architectural and decoration features, and consolidating the architrave and the upper part of the monument.

5   Impact of the CHUD for Sustainable Development

The actions carried out by the CHUD preserve cultural heritage sites that are cornerstones of a country's identity and history. In this case, they contributed to increasing the site's accessibility and its tourist appeal. Culture and heritage have been established as being amongst some of the most central components of a country's identity and attractiveness to visitors.

Cultural tourism in Lebanon lends itself to territorial development, as much as the Lebanese tourist industry is one of the central and most vitally important for the country's economy. Tourism in Lebanon is recognized as traditionally being one of the leading economic contributors to the country's GDP. Tourism in the country increased by 47% from 2020 to 2022, following the pandemic.

From the moment that the Italian Cooperation has been working on the best known sites in Lebanon, among which Baalbek holds the number one spot in terms of tourist numbers, these have come to represent an important income for the local area, as well as, more generally, for sustainable development.

From a social development perspective, the intervention has enhanced local human resources involved during the project, by carrying out local capacity building, training local experts, raising awareness and encouraging participatory processes. By involving local communities, including disadvantaged groups, the intervention has succeeded in building a shared vision of development. Different events involved local civil society, while authorities made sure to communicate and provide information on the activities, stressing that the preservation of the area's cultural heritage should involve everybody, since it is a powerful driving force for cultural tourism, and impacts several important sectors able to generate income. Over the course of the project, numerous experts, men and women, were involved in training; among them architects, archaeologists, engineers, and conservators, all of whom worked on site, dealing with the structural problems of the ancient monuments, cleaning parts that needed curative action, conserving architectural and decorative elements, and consolidating the architrave and the monument's upper section. Finally, a great number of unskilled local people benefitted from informal apprenticeships during the works.

These actions encouraged communication, and increased awareness that cultural heritage is to be considered a cornerstone of sustainable development.

The results achieved by the CHUD program are in line with the UN Global Indicator for Sustainable Development Goals, particularly that which acts *to make cities and human settlements inclusive, safe, resilient and sustainable, promoting inclusive and sustainable growth, and productive employment*. The program provided technical assistance and training to the DGU—Directorate General of Urban Planning and DGA—Directorate General of Antiquities, with the aim of improving cultural heritage management and promoting institutional strengthening. It also represented an important step in the implementation of Italian expertise, whose knowledge and experience in the field has been widely acknowledged among stakeholders.

The Italian commitment will have a positive impact on the country, by increasing sustainable and inclusive economic development opportunities in the areas where the projects take place. The results have been shown to lead to the successful and effective management of cultural heritage, by linking it with key stakeholders and thereby contributing to the revamping of the tourist industry, while setting up new and effective services that benefit both local communities and tourists.

Lastly, the restoration and conservation results have been evaluated by independent UNESCO experts, who looked into how the works were proceeding with regard to pre-conservation surfaces and treatment of the columns and decorative elements of the capitals, lintel and architrave. They also looked into the techniques adopted and the results obtained, and ended up expressing a favourable opinion of the work done. Experts concluded by saying that the Jupiter Temple has been restored inch by inch and that this intervention can be considered the best practice of restoration and conservation in the Mediterranean.

The Italian Agency for Development Cooperation in Jordan

Alessandra Blasi

Abstract In Jordan, the conservation and valorization of cultural heritage plays an important role in enhancing social cohesion and enriching understanding of the past, in that it addresses the identity and integrity of sites of cultural and historical interest, and works towards valorizing living heritage assets. The rehabilitation of artefacts and heritage sites, as well as strategies directed at the conservation of the urban fabric, also indirectly impact livelihoods within local communities. These actions result in increasingly appealing assets, which then attract potential visitors. This, in turn, allows for profitable, visitor-oriented activities that can boost the tourist industry. Such socio-economic connectivity provides a foundation upon which the quality of life of local residents can be improved. In the long term, it can lead to an increase in inclusivity, in the response capacity of women and children to vulnerability, the adaptability of investments and, ultimately, the development of sustainable tourism policies and strategies, based on changes in consumer preferences, which would work to bolster recovery.

Keywords Cultural livelihood · Cultural tourism · Resilient communities

1 Opportunities in the Tourism Sector in Jordan

Jordan's tourism industry made up nearly 13% of the country's GDP in 2019, bringing in 5.3 million visitors, JD 4.1 billion revenues, and over 1 million visitors to the archaeological site of Petra alone. Jordan ranked 6th among Arab countries and 64th among 117 countries in the latest "Travel and Tourism development Index (TTDI)"[1] issued by the World Economic Forum (WEF) in the fall of 2022. The sudden rebound in international tourism, which is close to reaching pre-pandemic levels, is expected to contribute greatly to Jordan's economic recovery alongside the

---

[1] https://www.weforum.org/reports/travel-and-tourism-development-index-2021.

A. Blasi (✉)
Italian Agency for Development Cooperation, Amman, Jordan
e-mail: Alessandra.blasi@aics.gov.it

© The Author(s) 2024
M. Loda and P. Abenante (eds.), *Cultural Heritage and Development in Fragile Contexts*, Research for Development, https://doi.org/10.1007/978-3-031-54816-1_15

service and export industries. Additionally, tourism is one the six sectors that make up the green economy (tourism, energy, agriculture, water, transport, and buildings) outlined in the strategic document "Economic Modernization Vision 2033" aimed at stimulating growth and advancing Jordan's transition towards an environmentally sustainable economy. As of 2022, tourism-related receipts contributed JOD 4.1236 billion to the Jordanian economy, bringing in 149% of the 2022 target, 114.5% of the 2023 target and 100.6% of the 2024 target, all of which were set in the "Jordan National Tourism Strategy 2021–2025 (JNTS) 2021–2025", marking a recovery rate of 100.4 and of 94% for the number of tourists who visited the country in 2022, both compared with 2019.

At the same time, with a population of just over 11 million in 2022,[2] Jordan is a country where unemployment remains consistently high (17.9% of the total labour force in 2022). As a major contributor to the Jordanian economy, tourism plays a crucial role in the government's efforts to increase the livelihood opportunities of local communities, and its effectiveness in so doing calls for numerous interventions, such as the provision of services and infrastructure, site rehabilitation, management plans, valorisation of both tangible and intangible cultural assets, and capacity building programmes directed at local authorities. In particular, as set out in the 2021–2025 JNTS, Heritage Protection, pertaining both to the enhancement and safeguarding of sites of historical and cultural interest, is considered a fundamental element in tourism strategy, and key to the growth and development of the industry.

2   AICS Amman Development Programme in Support of Local Communities Through Cultural Heritage

In Jordan, Italian support focuses on humanitarian responses to the regional and refugee crises, and sustainable development interventions, based on the national and strategic priorities identified in the MoU of 2017–2019 and the €235 million Subsidiary Cooperation Agreement (2021–2023) undersigned in May 2021.

The safeguarding of cultural heritage and its restoration in a country that is historically and archaeologically rich, constitute a central pillar in the bilateral collaboration between Italy and Jordan. In line with the Italian Cooperation's strategy on tangible and intangible cultural heritage protection, the Italian Agency for Development Cooperation (AICS) implements programs in Jordan, which focus on the relationship between culture, heritage, sustainability and tourism, as drivers for both economic growth and social empowerment.

In identifying interventions relating particularly to SDG 8.9 and SDG 11.4 in support of culture, AICS in Jordan prioritizes the valorisation of cultural assets and the safeguarding of the rich and diverse living heritage of the country to bolster cultural tourism, which is in turn a vector of productive investment expected to boost economic growth and reduce unemployment rates, especially for the most vulnerable.

---

[2] https://data.worldbank.org/country/jordan.

Having been part of the rehabilitation and valorisation projects implemented across many of the most significant sites, Italian-funded interventions call for local communities to engage more effectively in the protection of heritage sites, through sustainable practices and the possible adoption of alternative livelihoods. Socio-cultural assets are also given significant consideration by AICS in Jordan through cooperation initiatives. The valorisation of cultural heritage traditions, the regeneration of cities, the development of sustainable tourism and the re-use of heritage buildings, all reflect the potential to enhance social capital, reinforce a feeling of belonging and ensure good conservation practices, which might then benefit both cultural heritage and the local population in the long term. Based on their skills assessment, beneficiaries among the local community are often at the receiving end of income-generating activities through Cash for Work (CfW) schemes and on-the-job capacity building, aimed at protecting the attractiveness of tourist destinations and safeguarding the historical, cultural and landscape heritage which, as a by-product, fosters prosperity and social cohesion.

In the city of **Jerash**, in the North of Jordan, AICS supported the establishment of a "Regional Institute for Conservation and Restoration" (RICR) in an attempt to strengthen efforts by the Jordanian Ministry of Tourism and Antiquities (MoTA) and the Department of Antiquities (DoA) to conserve and restore the city's cultural heritage. This was done by providing specialist training in the conservation and restoration of cultural heritage, in order to increase the quality and capability of the industry's labour market. As part of the infrastructure part of the program started in 2019, the United Nations Office for Project Services (UNOPS), restored the building that currently hosts the RICR, and provided the Institute with specialist equipment. In 2023, the Italian University of Roma Tre launched the two-year specialist training courses addressed to the DoA staff and to future professionals in the region, tailored to the needs and desires of the local community, while taking the assessed skills gap into account. Trainees from the community are indeed the future professionals that will learn to apply their acquired practical experience and conservation abilities to the safeguarding of cultural assets, thus forging a sense of ownership of and accountability towards their shared heritage.

Southwest of Amman lies **Madaba**, the "city of mosaics," which is home to ancient ruins, churches and Byzantine and Umayyad mosaics, including the Map of Jerusalem and the Holy Land. Mosaics constitute the town's richest architectural heritage. To ensure preservation of such treasures, the University of Perugia, with the help of a grant from the Government of Italy through AICS, worked in collaboration with the local Department of Antiquities on the restoration of pavement mosaics and the maintenance of four archaeological sites in the city. In this case, too, the approach focused on ensuring sustainability of the restored sites, since decision-making processes and training involved local authorities and communities (Fig. 1).

Heading south, the program launched by AICS Amman in 2021 in the protected area of the **Shobak** castle, represents a best practice with regard to interventions that support local communities through the enhancement of cultural heritage areas. The two-part program is designed to heighten the resilience of people at risk of poverty among the urban and rural communities of the Shobak castle area, as well

Fig. 1 Local workers carrying out conservation activities on the mosaic floor located under the destroyed *Church* of the Prophet *Elias*, including replacement of the cement, consolidating the mosaic edges and mechanical cleaning. *Photo* Alessandra Blasi

as help reduce their increased socioeconomic vulnerability following the Covid-19 pandemic. As the interruption of tourist flows has impacted the main source of livelihood in the area, the ongoing activities aim to strengthen inclusive and sustainable development, and speed up the socio-economic recovery of local communities, through the enhancement of the archaeological site and the tourist trails. The additional restoration of small units around the castle in the Al-Jaya village—for the purpose of creating soon-to-be small businesses, such as bed and breakfasts, handicraft shops and other tourist-oriented establishments—will lead to enhanced urban landscapes with the potential to generate livelihood opportunities for local communities.

Through a collaborative approach based on the involvement of local residents, local authorities, and stakeholders, the initiative led to sustainable improvements throughout the Al-Jaya village, in line with the needs and aspirations of locals, while at the same time fostering community engagement in ensuring socially inclusive urban development (Fig. 2).

Lastly, in a milestone effort by the Italian Cooperation relating to cultural heritage in Jordan, a development program was launched in 2012 in the **Petra** Archaeological Park (PAP), a World Heritage site since 1985 that highly contributes to Jordan's overall tourist industry. With a total area of 264 km^2, Petra, the ancient capital of the Nabatean Kingdom, is also in constant demand for preservation and management interventions. Initiatives carried out in Petra in partnership with the UNESCO office

Fig. 2 Community members participating in a joint assessment aimed at selecting units suitable for restoration, at the Al-Jaya village, Shobak. *Photo* Alessandra Blasi

in Amman aimed to reduce the risk of landslides in the Siq, the main entrance to the ancient Nabatean city of Petra, as well as to provide capacity building and employment opportunities to Jordanian and Syrian youth, centred on cultural heritage preservation and risk prevention. The latest initiative funded by Italy in Petra and completed in 2023, concerned itself with the conservation of the Royal Tombs, a complex of Nabataean monuments comprising the Palace Tomb, which was built between the 1st and the second century AD. All the program's actions called for the participation of local communities, considered both actors and recipients of the activities. Jordanian and Syrian youth were both engaged in the preservation of the World Heritage Site and the implementation of landslide preventive measures in the Siq, through training in heritage conservation, mostly addressing graduate and undergraduate students, but also PDTRA staff training, as well as conservation and risk mitigation projects on the trails, aimed at the creation of immediate employment opportunities (Fig. 3).

Alongside the safeguarding of tangible heritage assets, AICS Amman has recently reached out to stakeholders and implementing partners on the relevance of including Intangible Cultural Heritage (ICH) in the site management strategies. So-called "living heritage", which encompasses folklore, spiritual beliefs, social practices, know-how and traditions, is intrinsic to local communities, while its valorisation ensures that emphasis is placed on accountability, cultural diversity and reinforcement of communities' social identity, as it represents the immaterial heritage which is passed down from generation to generation.

Fig. 3 Survey mission of international experts to evaluate the conservation status of the Palace Tomb and assess priority of interventions, including rehabilitation of the water channel and cleaning of cisterns carried out by workers from the local communities. *Photo* Alessandra Blasi

Site Risk Assessment Methods in Archaeological Built Environments: The Case Study of Shobak Castle in Jordan

Michele Nucciotti⑩, Fadi Bala'awi⑩, Mauro Sassu⑩, Mario Lucio Puppio⑩, and Fabio Candido⑩

Abstract A methodology for risk assessment has been developed, considering the main features of the archaeological site of Shobak Castle within the framework of AICS activities in Jordan. The site, as of the survey date, being an archaeological built environment, exhibits characteristics of a deteriorated urban setting with multiple needs and criticalities. Some areas have been subjected to archaeological excavation, temporary construction works, and subsequent surveys. To plan the rehabilitation works to be carried out by the University of Florence and to establish priorities for long-term management by the Department of Antiquities of Jordan, a comprehensive analysis of the entire Shobak Castle site was conducted. This analysis assessed different levels of risk, both for people and buildings, in accordance with the principles outlined in the Italian Consolidated Act on Safety (Legislative Decree no. 81 of April 9, 2008, and subsequent amendments). The results of this work, which

Each author had contributed to the present essay as follows: Michele Nucciotti, paragraph 1; Michele Nucciotti and Fadi Bala'awi, paragraph 2; Mario Lucio Puppio, Fabio Candido, and Mauro Sassu, paragraph 3; same as above for paragraph 4. The conclusions were written by all authors.

M. Nucciotti (✉)
University of Florence, Via San Gallo 10, 50129 Florence, Italy
e-mail: michele.nucciotti@unifi.it

F. Bala'awi
Department of Antiquities of Jordan, Jabal Amman - Third Circle - Abdul Moneim Al Rifai Street - Building No. 21, Amman, Jordan
e-mail: fadi.balaawi@doa.gov.jo

M. Sassu · M. L. Puppio
University of Cagliari, Via Marengo 2, 09123 Cagliari, Italy
e-mail: msassu@unica.it

M. L. Puppio
e-mail: mariol.puppio@unica.it

F. Candido
University of Pisa, Largo Lucio Lazzarino 1, 56121 Pisa, Italy
e-mail: fc@sundaymorning.it

229

have been documented in various worksheets, are summarized in the risk matrix and a series of safety maps presented herein.

Keywords Shobak castle · Site management · Archaeology · Site risk assessment

1 Shobak Castle and Its Cultural Values in the Frame of AICS Activities in Jordan

The archaeological site of Shobak Castle in Jordan (also spelled "Shawbak" in other publications), one of the most significant medieval fortified settlements in the Mediterranean region, has been the subject of study since 2002 by the archaeological mission "Medieval Petra—Shobak Project" of the University of Florence SAGAS Department (Vannini 2007). Such research has been conducted in collaboration with the Department of Antiquities of Jordan and in partnership with the Italian Ministry of Foreign Affairs and International Cooperation.

The historical and cultural value of the site is of great significance, both nationally and within the context of the Euro-Mediterranean medieval period (Vannini and Nucciotti 2009, 2012). The archaeological stratification at Shobak preserves traces of settlement dynamics and cultural facies spanning a broad chronology from ancient Roman times to the twentieth century. For this reason, Shobak Castle provides a privileged viewpoint on cultural transitions and entanglements of extraordinary interest, both for the scientific community and for non-scientific audiences, including residents and tourists. It serves as a true archaeological observatory, allowing one to traverse the history of the Mediterranean medieval period, from the Crusader era in the twelfth century to the Ayyubid and Mamluk periods from the twelfth to the sixteenth century, and through to the Ottoman era and the formation of the Hashemite Kingdom of Jordan in the twentieth century. Moreover, recent research conducted by the University of Florence on the nearby site of Jaya (Vannini 2020) and within the castle itself (in 2022, unpublished), as well as research conducted by IFPO in the nearby site of Dosaq (Imbert and Vigouroux 2020), are also revealing the site's significant value in the greater context of global histories and, more specifically, for the archaeology of medieval Eurasia. For instance, with reference to this last aspect, network analysis of imports at Shobak between the twelfth and fourth centuries, including materials from Syria and Egypt, as well as imports from China, Persia, and likely India (Fig. 1), highlights the prominent role played by the site in the long-distance global connectivity during the medieval period: a novelty that is of much interest for the site's interpretation and future development.

From a material perspective, Shobak is characterized by the presence of large architectural complexes from the Crusader, Ayyubid, and Mamluk eras that, despite being in a state of ruin, still retain significant structures: what we address in this article as a "built archaeological environment" (Fig. 2). This presents an undeniable

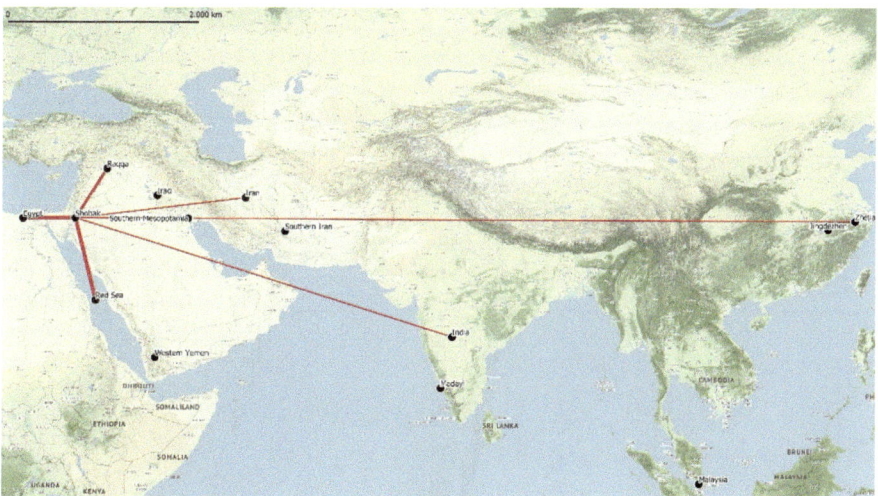

Fig. 1 Imports to Shobak Castle in the twelfth–fourteenth centuries (network analysis by Marco Moderato)

opportunity for the development of tourism at the site. To fully realize this potential, the Italian Agency for Development Cooperation (AICS) funded the project "Programme to support socio-economic recovery in the protected area of Shobak Castle" in 2021 (AID 012253/01/1), implemented by the University of Florence (project director Michele Nucciotti) and targeting needs expressed by the Department of Antiquities of Jordan (and more generally of interest to Jordan's Ministry of Tourism and Antiquities), primary beneficiary and planning partner for the intervention. The overarching goal of the AICS funded project is to support socio-economic recovery in the Shobak area of Jordan, promoting inclusive and sustainable local development through the rehabilitation and valorization of the site's tangible and intangible cultural assets. Within this framework, specific objective number 2 aims to "promote sustainable and participatory territorial development of the protected area and create employment opportunities for local communities". This includes the structural securing and musealization of a selection of medieval monuments at Shobak Castle and the design of new tourist paths, according to the expected result no. 2 "Reshaping of the site of Shobak Castle by securing a selection of monumental emergencies".

In this context, the development of the Risk Assessment methodology presented in this article integrates risk management with archaeological and architectural knowledge produced by research on the site (Nucciotti 2007; Nucciotti and Pruno 2016; Nucciotti and Fragai 2019) and public archaeology approaches aimed at breaking down intellectual barriers for tourist site interpretation (Nucciotti 2019). Moreover, the methodology works also as a mainframe of coordination between actions developed under the umbrella of AICS and those promoted and carried out at the site by the Department of Antiquities of Jordan. The methodology was therefore instrumental

Fig. 2 Panoramic view of Shobak Castle from the South (photo by Mauro Foli)

in building a closer collaboration and a profound sharing of practices and objectives between local authorities and international planners.

Based on this background, the article presents the step-by-step review of the process and results of risk assessment strategies adopted in Shobak, thought of as a replicable model to be considered for similar cases as well as for historic (urban) built environments at large.

2   Developing a Scalable and Replicable Risk Assessment Pipeline

From a practical standpoint, the implementation of the Risk Assessment plan involved a series of project steps, analysis, and knowledge transfer activities carried out over the ten months spanning from July 2022 to May 2023. These activities were closely coordinated between the University of Florence team and the technical team of the Department of Antiquities (DoA), under the supervision of the General Directorate of the Department of Antiquities of Jordan. Here is a summary of the key milestones in the pipeline:

1. Structural Vulnerability Mapping (July 2022): This task required both desktop research and on-site fieldwork. Desktop research involved collecting, organizing, and validating documents produced between 2006 and 2019 by the University of Florence archaeological mission to understand the site. This included surveys

and 3D photogrammetric models (Drap et al. 2009) created over two decades to assess major structural transformations and the speed and intensity of decay processes in historic buildings in Shobak. The result of this task was a preliminary mapping of vulnerabilities.

2. Evaluation and Prioritization of Vulnerabilities and Initial Proposal of Inter-vention Methodologies (July–October 2022): This task spanned approximately 3 months of fieldwork and aimed to acquire data on the actual presence of struc-tural instability and the activation of potential structural failure mechanisms in the facades of historic buildings in Shobak. Activities included the placement of fragile mortar links in the areas to be assessed (July 2022), with a follow-up assessment after three months, and assigning a severity/criticality index to each identified vulnerability (October 2022). Subsequently, potential intervention methods were selected for addressing or mitigating the identified vulnerabilities.

3. Sharing Objectives and Criteria for the Use of the Risk Assessment Tool between DoA and UniFi (November 2022): This task involved joint collaboration between the University of Florence team and the DoA team, including a thorough discus-sion of the results from the previous step, the selection of final intervention methods by the General Directorate of DoA (chosen from those proposed by UniFi in the previous task), and the specifics of expected usability of the Risk Assessment Plan. It was decided, for example, to consider the entire site for the Risk Assessment, to identify and manage safe tourist paths that cater to both visitors' needs and design requirements. Additionally, a numeric color matrix was associated with the Risk Assessment to facilitate its use by DoA technicians and to enhance the replicability of the adopted strategy in other archaeological sites.

4. Finalization of the Initial General Intervention Plan (December 2022): The data collected, and decisions made in the previous task allowed for the creation of an initial general intervention plan. This plan serves as a stable foundation for safety actions and (future) restoration efforts, with detailed information on methodolo-gies, materials, processes, and cost estimates. It should be considered a primary document in the present and future management strategies for the Shobak Castle.

5. Finalization of the Risk Assessment Plan (February 2023): Following the approval of the general intervention plan, in coordination with the DoA, the Risk Assessment Plan was finalized, along with the release of the risk matrix (see the following paragraphs for details). The final document aimed to produce a holistic site management tool, linking risk management closely with the management of access to the historical and cultural attractions that constitute the primary draw for tourists. Through the Risk Assessment Plan, the management can make informed decisions and evaluate the impact on tourist flows of structural safety measures.

6. Knowledge Transfer to DoA Personnel (May 2023): In May 2023, a dedicated training session was conducted for DoA operators on the Risk Assessment Plan. This task assessed the feasibility of effectively transferring complex expertise to a specialized audience and foreshadowed a positive impact at the national level in Jordan, increasing the DoA's capacity to map and manage structural (and tourism-related) vulnerabilities in Jordan's archaeological sites.

3   Methodology and Principles of Risk Assessment

The theoretical framework for risk assessment is based on the established Italian Testo Unico sulla Sicurezza—Consolidated Act on Safety (Testo Unico sulla Sicurezza 2008). Although originally designed for risk assessment and interferences on construction sites, it can be effectively adapted for managing safety at an archaeological site with urban character as Shobak Castle.

3.1   Terminology

Danger. The characteristic or intrinsic quality of a specific factor with the potential to cause damage.

* Damage cause or origin (UNI 11230 2007);
* Potential source of damage.

The danger is an intrinsic property (of a specific situation, object, substance, etc.) unrelated to external factors. It is a situation, object, substance, etc. that because of its characteristics can create damage.

Damage

* Any negative consequence descending from the occurrence of an event (Testo Unico sulla Sicurezza 2008);
* Physical injury or health damage;
* Seriousness of the consequences occurring with a danger happening.

The magnitude of consequences (M) can be expressed as a function of the number of involved subjects in that specific danger with that damage level suffered.

Risk. Probability to reach the potential level of damage when exposed to a specific factor or agent to their combination. Risk is a probabilistic concept—it is the probability that an event occurs causing damage to people or things. The notion of risk implies the existence of a source of danger and the possibility that this source becomes damage.

Prevention. Any necessary actions—given the specific work, experience, technique, and situation—needed to avoid or reduce risks with respect for people's health and the integrity of the environment.

Prevention measures are both structural or organizational, as:

* Information, formations, workers and visitors training;
* Planning, construction, and right use of spaces, structures, tools, machines, and systems;
* Avoiding dangerous situations that could cause possible danger (risk);
* Adopting adequate behaviors and procedures.

Protection. Safeguard against anything able to cause damage. Element intercutting between someone or something susceptible to suffer damage and the cause of damage.

3.2 Safety Risk Identification for Shobak Castle Archaeological Site

The Italian Consolidated Act on Safety is dedicated mainly to the safety of workers. The Shobak Castle's Risk Assessment was adapted to adhere to the requirements expressed by the Jordanian Directorate of Antiquities and of the AICS funded project, to be dedicated both to the safety of people—visitors and workers as tourist guides, keepers, etc.—and building with archaeological features.

Safety risk factors for people (visitors, tourist guides, etc.):

- Workplaces in enclosed spaces;
- Visit areas in enclosed spaces;
- Enclosed spaces in general;
- Workplaces and paths through ruins;
- Workplaces and paths on higher heights;
- Unprotected paths and workplaces;
- Uneven paths and areas.

Safety risk factors for buildings and construction:

- Anthropic actions (mainly caused by visitors);
- Bad maintenance;
- Weathering;
- Poor construction techniques.

3.3 Risk Estimation

Risk estimation is the identification of the possible seriousness of damage and the probability of its occurring. Risk can be expressed as a function of probability and magnitude:

$$R = (F, M) \tag{1}$$

R = Risk

F = Frequency or probability of occurring consequences

M = Magnitude (seriousness) of the consequences (damage to people or buildings).

Table 1 Criteria for probability estimation

Probability value	Level	Criteria
1	Improbable	Unknown occurring The occurrence of damage will cause incredulity
2	Less probable	Very few occurring are known The occurrence of damage will cause a big surprise
3	Probable	Some episodes are known The occurrence of damage will cause some surprise
4	Very probable	Episodes in similar situations are known The occurrence of damage will not cause a surprise

To define F and M it is possible to use two scales made of four values, each one corresponding to a specific level of possibility and a specific seriousness of damage, as described in Table 1.

3.4 Numerical Evaluation of Risk R

The risk matrix indicated in Fig. 3 has been used.

The numerical evaluation of risk R implicates the fulfilment of preventative and protection measures related to the risk evaluation (Table 2).

Fig. 3 Matrix of risk

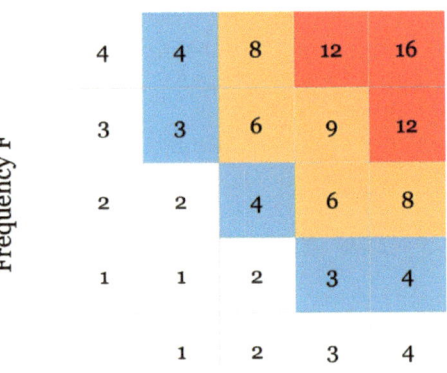

Table 2 Prevention and protection measures for risk reduction

General safeguard measures			
Technical measures	Management measures	Procedural measures	Personal protection measures
Improvement of technical tasks, systems, and structures	Management of work processes	Emergency plans and first response emergency aid	Personal protection
Improvement of visit spaces and workplaces	Management of visit flows	Restoration and refurbishment plans and schedules	Collective protection
Replacement or refurbishment of dangerous elements	Management of visit behavior	Control and prevention procedures	
Monitoring systems	Information, formation, and training for workers and visitors		

3.5 Risk Reduction

Any corrective actions should reduce the risk until:

Tolerable risk: Risk accepted after risk evaluation. The tolerable risk is also called acceptable risk. The tolerable risk should not require any further corrections.

Residual risk: Risk that remains after the risk treatment. The residual risk includes also not recognizable risks (UNI 11230 2007).

$$R = Pr \times D \tag{2}$$

R = Risk

Pr = Prevention (reduces the occurring probability) − Protection (reduces the level of damage)

D = Damage.

3.6 Risk Revaluation Process and Tools

The risk evaluations are to be intended as an iterative process as shown in Figs. 4 and 5.

The following tools have been used by the authors for the current risk evaluation.

Surveys and interviews notes:

- Visits with people who work in the specific place;
- Photographic survey;

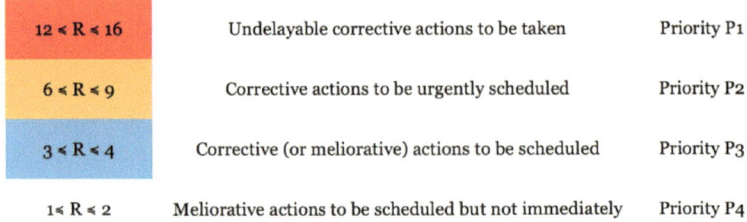

Fig. 4 Criteria for risk evaluation and priority of intervention

- Interviews with the workers and visitors;
- Interviews with managers and accountables.

Layouts and technical plans

- Spaces management;
- Emergency paths and exits;
- Planned systems installation;
- Recognition of areas with specific risks;
- Interferences.

However, risk management involves continuous work of monitoring and safety management implementation that should be carried out by the site managers. The following tools should be constantly implemented:

Hierarchy of safety managers and responsible

- Check of organization to implement appropriate management of safety measures;
- Monitoring;
- Monitoring of the preservation state of the building;
- Monitoring of the visitors' behavior.

Statistical analysis of damages, accidents, and missed accidents

- Analysis of how the site management handled an occurred injury or accident;
- Research for statistical data for suggestions to risk management.

3.7 General Safeguard Measures

Following risk evaluation, prevention, and protection measures are set up to be adopted for risk reduction. These measures, illustrated in Table 2, can be grouped as:

Technical measures

Actions on visiting spaces, workplaces, and systems. These measures can be preventive or aimed at reduction or limiting risk.

Management measures

Actions aimed to improve the performances of entropic factors (both visitors and workers).

Procedural measures

Actions to improve behaviors and activities inside the site (both for visitors and workers). These can be rules, improvements and updates, internal procedures, etc.

Personal protection measures

Personal protection devices (PPD)

Collective protection devices (CPD).

4   Risk Assessment

The risk assessment of an archaeological site should consider its special features concerning risks to people and historical monumental works. Some recurrent pathologies of historic masonry works are presented in Puppio et al. (2023) or also in Puppio et al. (2021) that deal with the significant case of the historical Urban walls of Volterra in Italy. Other significant contribution based on the analysis of UNESCO sites are (Sassu et al. 2013, 2017). The site is divided into homogeneous areas. Risks for people are analyzed by considering the effective hazards that may occur in each area according to its actual condition.

4.1   Shobak Castle Site Organization

The site was organized in 12 quarters/homogeneous-areas, to permit a good management of different areas. The quarters partition is based on the following criteria:

- topology of the site;
- visit's features or archaeological areas;
- homogeneity of risk factors;
- cultural and tourist values.

Each quarter contains several risk evaluation sheets associated with the risk factors evaluation, for any specific location or feature. The risk evaluation sheets are the core of this work. They are intended as a flexible tool to be implemented by site managers. Following updated risk evaluations in the future, the sheets can be possibly increased in number or grouped. The sheets should be updated regularly following the criteria stated in the previous chapters. Also, the quarters' subdivision can be modified in case of need, following the same criteria.

4.2 Risk Factors

The risk factors were established in relation to two main groups:

- people—visitors, archaeologists, tour guides, keepers, etc.
- buildings—structures, archaeological features, etc.

The risk for people does not consider safety assessments for workers on the construction site (workers involved in restorations, reparations, maintenance, etc.) because generally these evaluations are fulfilled by site managers or health inspectors and should be in specific assessments under Italian regulations, for instance, a document called Piano Operativo di Sicurezza—POS "Operational Safety Plan" should be prepared for that specific purpose. It is a project plan of the site works concerning safety issues. Based on scientific literature, experience, and site knowledge, the following risk factors were established as the most frequent on the site:

Risk for the people

- fall of materials from above;
- fall on ground;
- fall from above;
- microclimate.

Risk for the structures

- fall/drop of materials;
- local or global instability of structural elements;
- washout/runoff;
- anthropic stresses/impacts.

The numerical risk evaluation R is associated to a priority of intervention, as stated in the previous chapters, and to a coded color, following the risk level and priority. This color is used as a visual reference in the site map, as shown in Fig. 3, to give a current overview of the mapped risks.

Also, a touristic and cultural value is associated with each feature in any quarter, to give a deeper understanding of the actions and interventions required and in order for local tourism managers to anticipate effects on touristic attractivity of the site of the prioritized safety interventions.

4.3 Sheets Anatomy

The risk evaluation sheets are the core of the risk assessment work. As specified above, they are intended as a flexible tool to be continuously implemented by site managers. Following updated risk evaluations in the near future or risks reduction actions, the sheets can and should be potentially increased in number, or grouped. The sheets should be updated regularly following the criteria stated in the previous chapters. Also, the proposed quarters' subdivision can be changed following the same criteria. The site plans offer an overview of the current state-of-risk on the site and should be used both as a visual reference and management tool. The sheets, for any quadrant, are organized in the following parts:

- Cultural and touristic values in the quadrant;
- Registry section with quadrant of belonging, specific feature identification, description, main risk factors identification;
- Photographic reference;
- Risk evaluation for the people;
- Risk evaluation for the structures;
- Assessment on risk ranking with recommended actions for the reduction of the main risk factors.

The recommendation of this framework is traduced in graphical instruction for a direct use as shown in Figs. 5 and 6. Figure 6 shows the division of the archaeological site into homogeneous zones within which the risk analyses are carried out (including a list of areas to be forbidden to tourists due to the risk assessment Fig. 7).

5 Conclusion and Perspectives

The site risk-assessment methods developed for Shobak Castle highlight a safety model intended to serve as an interactive and adaptive tool. It is now focused on intervention prioritization and designed to be a flexible instrument capable of accommodating various on-site work scenarios. It will be reutilized as site retrofit activities continue to evolve.

The comprehensive management approach applied to the archaeological site of Shobak, as delineated in the proposed methodology, encompasses multifaceted considerations. These encompass the preservation of the site's physical fabric, the facilitation of visitor activities, and the preservation of its historical and structural heritage. The imperative to ensure the safety of tourists and visitors during their excursions, as well as the safety of scientists and laborers engaged in excavation and restoration efforts, must be seamlessly integrated with the optimization of the site's utility.

Moreover, the recognition of the historical significance inherent in these features should be coupled with the promotion of collaborative engagements among institutional managers, researchers, local communities, and tourists. This collaborative endeavor is undertaken with a steadfast commitment to the site's enduring preservation and valorization. Shobak Castle, as a built archaeological environment, serves as a paradigmatic case study for the development of scalable and replicable models of holistic management, which encompass elements of risk prevention and mitigation. These models are envisioned to find utility not only within the context of Shobak but also in other archaeological sites across Jordan and beyond. Through these efforts,

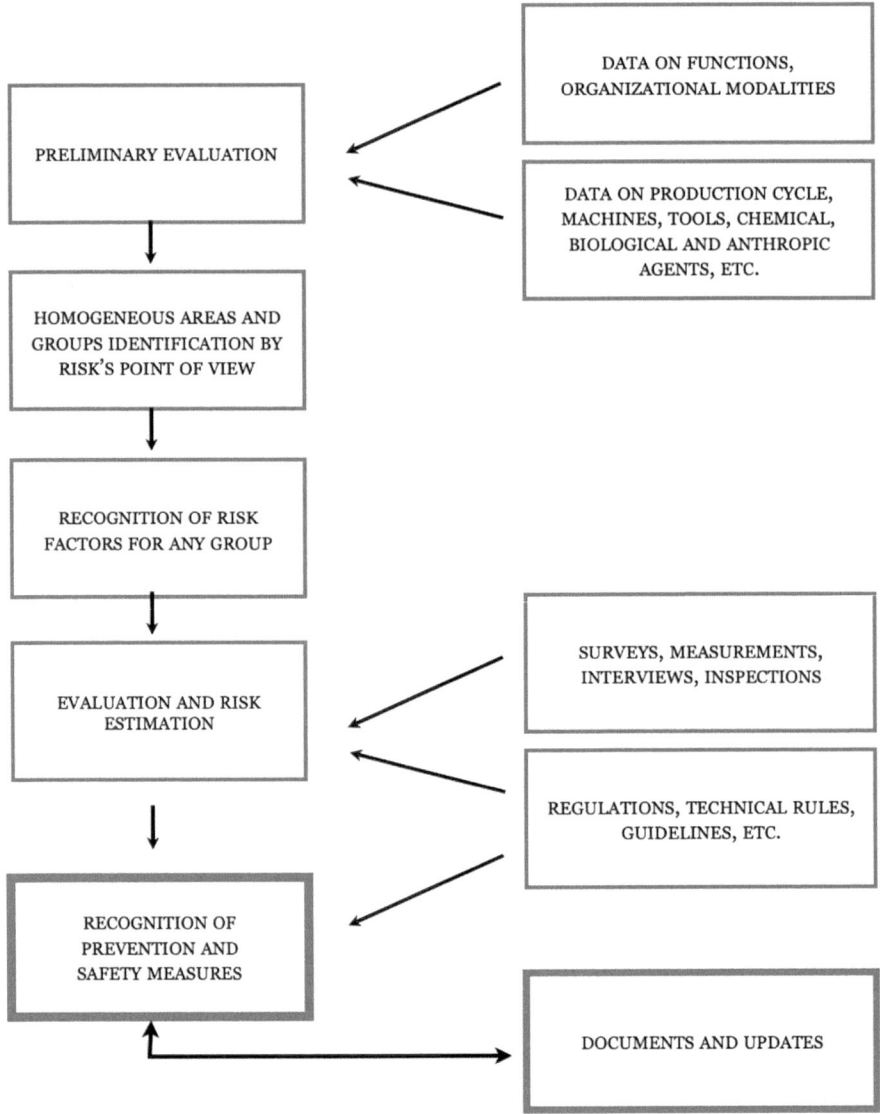

Fig. 5 Iterative procedure for risk assessment

the site is poised to fully realize its potential as a global "memory locus", fostering connections and dialogues among the local community, scholars, and visitors alike.

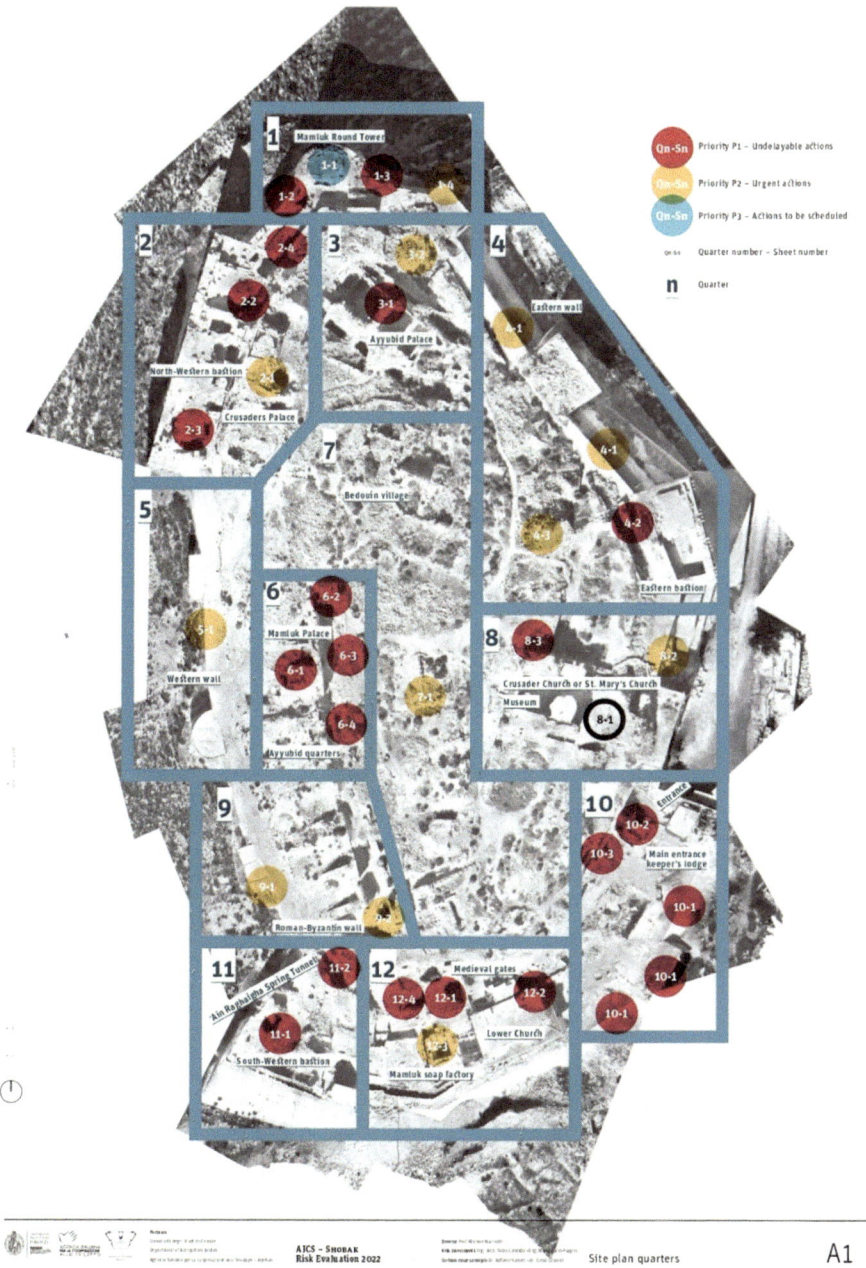

Fig. 6 Risk map of archeological site of Shobak Castle with the partition of the site and the priority of intervention

Fig. 7 Risk map of archeological site of Shobak Castle with forbidden area according on the safety assessment framework

References

Drap P, Seinturier J, Chambelland JC, Gaillard G, Pires H, Nucciotti M, Pruno E, Vannini G (2009) Going to Shawbak (Jordan) and getting the data back: toward a 3D GIS dedicated to medieval archaeology. In: 3rd ISPRS international workshop 3D-ARCH: "3D Virtual Reconstruction and Visualization of Complex Architectures", vol XXXVIII-5/W1. ISPRS

Imbert F, Vigouroux E (2020) Les inscriptions ayyoubides de Khirbat al-Dūsaq: histoire d'un puzzle épigraphique. Bulletin d'études orientales 2020/1(67-2020/21):111–131

Nucciotti M (2007) Analisi stratigrafiche degli elevati: primi risultati. In: Vannini G (ed) Archeologia dell'insediamento crociato-ayyubide in Transgiordania. Il progetto Shawbak. Insegna del Giglio, Florence, Italy, pp 27–55

Nucciotti M (2019) Archeologia Pubblica, distretti turistici e nuove storie rurali. In: Nucciotti M, Bo-nacchi C, Molducci C (eds) Archeologia Pubblica in Italia. Firenze University Press, Florence, Italy, pp 223–240

Nucciotti M, Fragai L (2019) Ayyubid reception halls in southern Jordan: towards a 'Light Archaeology' of political powers. In: 13th studies in the history and archaeology in Jordan. Department of Antiquities of Jordan, Amman, pp 489–501

Nucciotti M, Pruno E (2016) Great and little traditions in medieval Petra and Shawbak: contextualizing local building industry and pottery production in cc. 12–13. Archeologia Medievale (43):309–320

Puppio ML, Vagaggini E, Giresini L, Sassu M (2021) Landslide analysis of historical urban walls: case study of Volterra, Italy. J Perform Constr Facil 35(6). https://doi.org/10.1061/(ASCE)CF.1943-5509.0001647

Puppio ML, Sassu M, Safabkhsh A (2023) Damage and restoration of historical urban walls: literature review and case of studies. Frattura ed Integrità Strutturale 65:194–207

Sassu M, Andreini M, Casapulla C, De Falco A (2013) Archaeological consolidation of UNESCO masonry structures in Oman: the Sumhuram citadel of Khor Rori and the Al Balid Fortress. Int J Arch Herit 7(4):339–374

Sassu M, Zarins J, Giresini L, Newton L (2017) The 'Triple R' approach on the restoration of archaeological dry stone city walls: procedures and application to a UNESCO world heritage site in Oman. Conserv Manag Archaeol Sites 19(2):106–125

Testo Unico sulla Sicurezza, Legislative Decree no. 81 of April 9, 2008

UNI 11230:2007—risk management—terminology

UNI EN ISO 12100-1—machine safety—fundamental concepts, general design principles—part 1: terminology, methodology

Vannini G (ed) (2007) Archeologia dell'insediamento crociato-ayyubide in Transgiordania. Il progetto Shawbak. Insegna del Giglio, Florence, Italy

Vannini G (2020) Al-Jaya Palace and the New Shawbak Town. A Medieval frontier and the return of the urbanism in the Southern Transjordan. Stud Anc Art Civ 83–98

Vannini G, Nucciotti M (eds) (2009) From Petra to Shawbak, archaeology of a frontier. Giunti, Florence, Italy

Vannini G, Nucciotti M (eds) (2012) La Transgiordania nei secoli XII–XIII e le 'frontiere' del Mediter-raneo medievale. BAR Publishing, Oxford

The Role of Heritage Education in Post-conflict Livelihoods: Lessons from Mosul, Iraq

Georges Khawam and **Rohit Jigyasu**

Abstract This chapter explores the transformative role of heritage education in post-conflict livelihoods, focusing on Mosul, Iraq. It discusses the evolving United Nations definitions of "livelihoods" and the increasing recognition of culture and creative industries in post-conflict livelihoods. It highlights the role of intergovernmental organisations in implementing heritage-driven economic revival strategies, focusing on ICCROM Heritage Recovery Programme in Mosul implemented in the framework of the UNESCO "Revive the Spirit of Mosul" initiative. These strategies aim to generate livelihoods through tourism, cultural industries, and traditional craftsmanship, contributing to long-term stability and resilience in post-conflict regions. This chapter underscores the importance of heritage capacity building in increasing livelihoods after conflicts, as demonstrated by the preliminary success of the Mosul initiative, and concludes by emphasising the need for an integrated approach to heritage conservation and sustainable economic growth in post-conflict contexts.

Keywords Post-conflict heritage recovery · Capacity-building · Livelihoods

1 The Evolving Definition of Livelihoods

The United Nations (UN) definitions of "livelihoods" have evolved significantly since their introduction in the 1990s to refer to maintaining quality of life, followed in 1992 by the emergence of "sustainable development." Since then, the UN has considered the skills, resources, and activities people use to meet their needs and improve their well-being. It has enhanced its holistic understanding, progressively emphasising an integrated approach to livelihoods that recognises impoverished people's diverse livelihood systems and subsequent needs.

UN and Intergovernmental Organisation (IGO) publications underscore the role of culture and creative industries in post-conflict livelihoods, promoting sustainable development, peace, reconciliation, and resilience in conflict-affected societies. The

G. Khawam · R. Jigyasu (✉)
ICCROM Heritage Recovery Programme in Mosul, Rome, Italy
e-mail: rohit.jigyasu@iccrom.org

© The Author(s) 2024

M. Loda and P. Abenante (eds.), *Cultural Heritage and Development in Fragile Contexts*, Research for Development, https://doi.org/10.1007/978-3-031-54816-1_17

United Nations Development Programme (UNDP) report links crisis-related livelihoods, such as emergency employment, through their positive impact on income and well-being to long-term peace and stability. In 2017, the International Labour Organisation (ILO) published a guide to advocate for implementing market-oriented approaches in refugee livelihood programmes to facilitate their economic inclusion engagement with host communities. In 2018 and 2020, The United Nations Educational, Scientific and Cultural Organization (UNESCO) published the "Culture in City Reconstruction and Recovery (CURE) Framework" and the "Culture in Crisis: Policy Guide for a Resilient Creative Sector," focusing on culture in city reconstruction and recovery, promoting urban economic growth, inclusivity, resilience, and sustainability, and the second aiming to strengthen cultural and creative industries' resilience to contribute to the livelihoods of artists and cultural professionals. In 2021, the International Centre for the Study of the Preservation and Restoration of Cultural Property (ICCROM) published the PATH—Peacebuilding Assessment Tool for Heritage Recovery and Rehabilitation, which evaluates heritage's role in peacebuilding and provides guidance on designing conflict-sensitive heritage livelihood-related projects. UNCTAD's "Creative Economy Outlook 2022" report analyses the creative economy's current and future state and how it can help achieve sustainable development. The 2022 "UN-Habitat Urban Recovery Framework (URF)" emphasises crisis-sensitive employment and livelihood initiatives for urban recovery. The 2022 "ICCROM Analysis of Case Studies in Recovery and Reconstruction" briefly mentions livelihoods as a factor in holistic heritage recovery.

The emphasis across these documents is on leveraging heritage as a cornerstone for economic recovery. They underscore the pivotal role of heritage in generating livelihoods after conflicts by promoting heritage-driven economic revival, including tourism, cultural industries, and traditional craftsmanship. These efforts empower local communities and contribute to long-term stability and resilience, aligning heritage conservation with sustainable economic growth in post-conflict regions.

2   Heritage as a Generator of Livelihoods

In response, UN organisations and international NGOs began implementing projects reflecting this shift in post-conflict situations, supporting and increasing livelihoods in Syria, Yemen, Iraq, and Jordan through cultural heritage development projects, the most notable of which is the UNESCO flagship initiative "Revive the Spirit of Mosul," which all demonstrate the potential of cultural heritage to enhance livelihoods, contributing to sustainable development, social inclusion, and resilience. In the context of Syria, Yemen, Mosul in Iraq, and Mali, ICCROM has played a significant role in capacity building, training, and providing guidelines for protecting and restoring cultural heritage. These efforts have not only contributed to the preservation of cultural heritage but also to the improvement of livelihoods, social cohesion, and resilience in these regions. UNESCO's work in culture to improve livelihoods is multifaceted and spans various global areas. The organisation's initiatives primarily focus on leveraging cultural heritage to create employment opportunities, foster social cohesion, and promote sustainable development. One of the key projects is in Jordan, where UNESCO has created employment opportunities and provided skills training for 283 Syrian refugees and vulnerable Jordanians, addressing unemployment and boosting tourism in northern Jordan.

Similarly, in Yemen, UNESCO launched a project in 2018 that employs young Yemenis in heritage restoration and preservation while promoting social cohesion and resilience through cultural programming, surveying over 8000 historic buildings and 151 buildings, and enrolling over 2500 young workers (UNESCO 2023). Economic initiatives linked to living heritage that are established by UNESCO have the capacity to generate revenue for the communities, groups, and individuals impacted, in addition to enhancing livelihoods and decent employment opportunities in the local economy (UNESCO n.d.). It has been working on integrating conservation, sustainable tourism, and local livelihood opportunities. These efforts aim to improve communities' livelihoods while preserving and promoting their cultural heritage. By integrating conservation and sustainable tourism practices, UNESCO understands that it can create a balance between preserving cultural heritage and boosting the local economy. By generating income through economic activities linked with living heritage, communities, groups, and individuals are empowered to improve their livelihoods. Through these efforts, UNESCO aims not only to preserve and promote cultural heritage but also to contribute to the overall development and well-being of the communities involved.

On the other hand, ICCROM is a unique intergovernmental organisation with an inspiring global mission to protect cultural heritage worldwide, and its individual mandate, expertise, and international outlook have made it essential to preserving our past since its creation in 1956. Its work in post-conflict recovery focuses on preserving and restoring cultural heritage, which is essential for rebuilding communities and promoting social cohesion. ICCROM and various other organisations play a crucial role in supporting sustainable livelihoods within communities that embrace cultural heritage. By collaborating with local communities, ICCROM helps develop strategies and initiatives that promote the conservation and management of cultural heritage while simultaneously fostering economic growth. ICCROM equips individuals with the necessary skills and knowledge to engage in sustainable tourism practices through training programmes and capacity-building workshops. This empowers local communities to not only preserve their cultural heritage but also create economic opportunities that enhance their livelihoods and overall well-being.

ICCROM work includes workshops and training programmes on post-conflict recovery that aim to exchange ideas and experiences, case studies, and classroom activities on assessment, design, and recovery strategies. ICCROM collaborates with civil society organisations to improve people's living conditions, address climate and environmental change, and promote sustainable development. They also make guides and publications about cultural heritage in post-war recovery, like the Cultural Heritage in Post-War Recovery guide, which stresses how important cultural heritage is in recovering from armed conflict and how it needs to be planned into recovery after a conflict. Lastly, ICCROM promotes an integrated and multi-hazard risk management approach to heritage conservation.

ICCROM's post-conflict recovery programmes generally focus on restoring and protecting cultural heritage, which is an essential part of rebuilding communities and bringing people together. ICCROM collaborates with civil society organisations, organises symposiums, and develops tools and guidelines to promote sustainable development, address climate and environmental change, and enhance living conditions.

3   The UNESCO "Revive the Spirit of Mosul" Initiative

The "Revive the Spirit of Mosul" project, launched by UNESCO and supported by ICCROM, is a testament to the transformative power of heritage capacity building in post-conflict environments, thanks to financial support from the Government of the United Arab Emirates and the European Union. The project's approach goes beyond merely providing employment; it fosters on-the-job learning opportunities, equipping professionals and workers with the necessary skills to contribute to the reconstruction efforts. The importance of heritage capacity building in increasing livelihoods after conflict cannot be overstated. This is particularly evident in the context of this project, which has been instrumental in the recovery and rehabilitation of Mosul, Iraq, following years of conflict. The project's approach goes beyond simply providing employment; it also fosters on-the-job learning opportunities, ensuring the active participation of residents in the reconstruction efforts. A component of this initiative was the extensive on-site training that ICCROM designed for Mosul to successfully train young building professionals and revive traditional building crafts to conserve and restore historical monuments in Mosul in its ongoing post-conflict reconstruction phase. This initiative aims to restore and rebuild the cultural heritage of Mosul, Iraq, which was severely damaged and destroyed during the conflict with ISIS and actively involves collaborations with local communities, experts, and international organisations to ensure the preservation of historical sites, such as the Al-Nouri Mosque and the Al-Hadba Minaret, while also providing opportunities for social and economic development in the region.

4 The ICCROM Heritage Recovery Programme in Mosul

4.1 Overview

ICCROM designed the course in close collaboration with local stakeholders as a comprehensive capacity-building programme tailored to the specific needs in Mosul and delivered in two 9 month tracks for 50 young professionals such as architects and engineers, as well as 79 semi-skilled and skilled artisans in building crafts such as alabaster work and blacksmithing that give the city its unique character. The training methodology combined theoretical learning with practical site work and workshops. By equipping participants with specialised skills and knowledge, the HRP aimed to facilitate their integration into Mosul's labour market, supporting heritage recovery and rehabilitation.

4.2 Building Capacity for Professional Development (Track 1)

Track 1 was designed to develop the expertise of 50 young architects and civil engineers throughout two training cycles, each lasting for 9 months and involving 25 participants. The main objectives of this track are to introduce participants to heritage management and conservation, train them in planning and implementing holistic heritage recovery processes, and equip them with the technical competencies and soft skills needed to participate in heritage recovery and reconstruction initiatives in Mosul. The second cycle was revised during a meeting at the Headquarters of ICCROM in Rome with the Course Team and core resource people to better adapt the curriculum to the needs of the Moslawis after the experience and feedback gathered throughout the first cycle. Figure 1 explains the flow in interrelations between the seven modules that constitute each cycle of Track 1:

1. **Orientation**: This module introduced the basic principles of urban heritage conservation, both tangible and intangible. It addressed heritage values and current heritage management and conservation approaches, focusing significantly on the Moslawi heritage. The goal was to understand Mosul's complex and multifaceted heritage and the vocabulary used to describe cultural heritage.
2. **Situation and Context Analysis**: This module aimed to teach participants how to map and establish the context in a post-crisis environment, focusing on heritage recovery. It covered methods for conducting a situation and context analysis, identifying and mapping stakeholders, and addressing issues from security risk assessment to legal and institutional framework analysis.
3. **Documentation**: This module provided knowledge of principles and techniques for documenting and surveying cultural heritage at various levels. It included analysis at the regional/city level to identify various historical layers of the

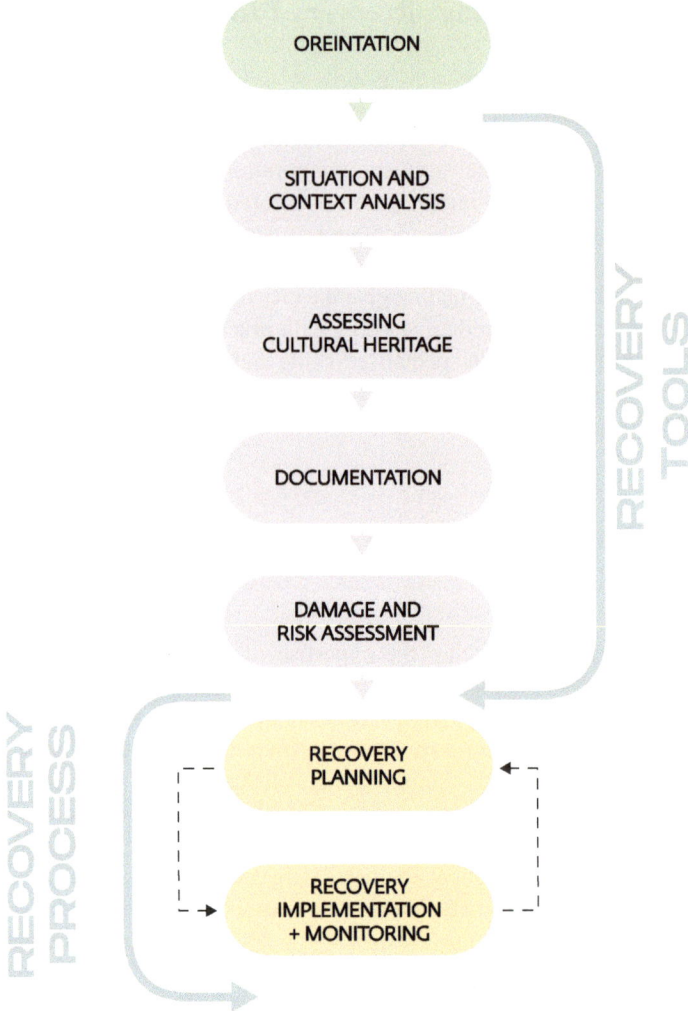

Fig. 1 Revised diagram of track 1. © ICCROM

city, tracing remains of the historic fabric, traditional infrastructure, and urban patterns.

4. **Assessing Cultural Heritage Significance and Values**: This module provided an understanding of the methodology for assessing cultural heritage significance and values at city, neighbourhood, building, or component levels. The focus was on understanding the heritage attributes and associated values and their prioritisation to the recovery process.

5. **Damage and Risk Assessment**: This module concerned the identification of damages to historic buildings in the context of Iraq and, more precisely, Mosul

Fig. 2 Field exercise on damage assessment. © ICCROM

City. It involved the development of the methodology for conducting damage assessment for cultural heritage at several levels and analysing and assessing different types of hazards and the elements at risk.

6. **Recovery Planning**: This module introduced the main aspects of recovery planning for Mosul from a management perspective. Topics included the broader impact of planned recovery actions on the historic urban fabric, planning principles for recovery, preparation of a Master Plan for Recovery, community and stakeholder engagement, and financial aspects of recovery.

7. **Implementation and Monitoring**: This module focused on the procedural steps and considerations for implementing micro-level planned recovery interventions. Topics included primary intervention considerations, stabilisation and security measures, impact assessment, implementing project-level interventions, and engaging community and other stakeholders in recovery implementation (Figs. 2, 3, 4, 5, 6 and 7).

4.3   Building Crafts Revival and Upgrading (Track 2)

The second track, Building Crafts Revival and Upgrading, was developed to "build back better" and allowed semi-skilled craftspeople to enhance their skills while contributing to restoration efforts. Based on the results of an assessment of crafts and craftspeople in Mosul, four traditional crafts were identified as needing priority recovery: alabaster work, stonemasonry, carpentry, and blacksmithing, and involved international experts in bringing back essential know-how that could have been lost as the industry favoured less expensive techniques and materials.

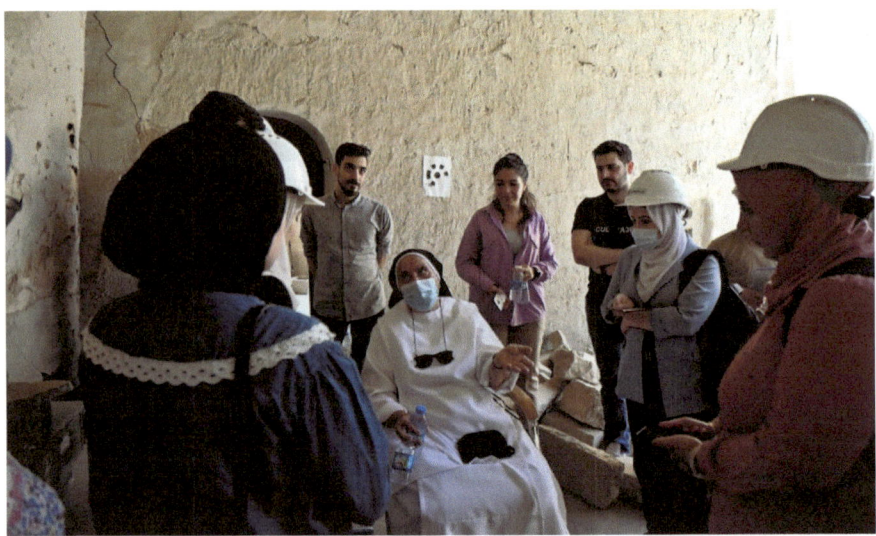

Fig. 3 Interview with stakeholder. © ICCROM

Fig. 4 Value assessment fieldwork. © ICCROM

Fig. 5 Classroom studio work. © ICCROM

Fig. 6 Fieldwork practical exercise. © ICCROM

Fig. 7 Field visit to the top of the Al-Minaret. © ICCROM

A needs assessment mapped the existing typology of crafts, the level of skills available in Mosul, the locations of craftspeople, their demographic, social and economic profiles, the impact of conflict on these crafts, and current challenges and needs. It helped identify the master craftspeople who led the workshops. The training takes into consideration affordability as well as the ground realities (ICCROM 2021b). To ensure sustainability, the programme establishes links with the building industry to guarantee the availability of traditional materials and, as much as possible, focuses on recycling usable materials from the rubble (ICCROM 2021a). Finally, for the knowledge and skills imparted to both professionals and craftspeople to be economically beneficial, the programme design was informed by thorough market research on the business opportunities and challenges of built heritage professionals and craftspeople with the hope to contribute to a truly holistic, sustainable, and resilient heritage recovery of Mosul.

The carpentry workshop took place over 51 days and covered personal safety; different types, sizes, shapes and origins of wood; measurements; wood cutting and basic joinery techniques; and producing and repairing doors, windows and furniture using modern and traditional techniques. The 66 day alabaster workshop focused on building techniques, carving and sculpting. Supported by local and international experts, trainees explored restoration using epoxy and various types of mortar. The 59 day workshop on stone masonry included plastering and construction of walls, arches, vaults and domes, including theoretical discussions on the characteristics of Mosul limestone to the tools, cleaning and practical sessions working with stone itself. Blacksmithing required additional resources as the need assessment revealed that hand-forged handrails and window protection without modern electric welding had been lost. Because similar techniques are present in other historic Arab cities, ICCROM involved a skilled master blacksmith from Cairo who undertook several missions to Mosul and initially conducted a 15 day Training of Trainers, a 52 day workshop for semi-skilled blacksmiths, and a second stage (ICCROM 2023) (Figs. 8, 9, 10 and 11).

Fig. 8 Female participants during the alabaster workshop. © ICCROM

Fig. 9 Building a traditional dome during the stone masonry workshop. © ICCROM

Fig. 10 Door production during the carpentry workshop. © ICCROM

Fig. 11 Training of trainers of master blacksmiths. © ICCROM

5 Impact on Livelihoods and Economic Recovery in Mosul

The ICCROM Heritage Recovery Programme provided its graduates with the necessary skills to integrate into the labour force active in the reconstruction of Old Mosul in response to the growing labour market demand, enabling them to earn a living and contribute to the local economy. On one hand, this involved training architects and engineers in adapting heritage buildings and sites to new uses that align with the local community's needs, and on the other, training Track 2 participants on the use of hybrid traditional and modern techniques and materials, increasing the sustainability and affordability of the heritage recovery process. In parallel, ICCROM raised the awareness of local contractors on the importance of using traditional materials in the recovery process and its cost-efficiency when coupled with hiring trained craftspeople.

Upon graduation, to encourage continued learning and facilitate practice, each Track 1 participant received a collection of seminal reference books in the field of heritage conservation. On the other hand, Track 2 participants received a start-up kit of essential tools for each craft to encourage practising the acquired skills. Further enhancing its action's sustainability, ICCROM hired several graduates of both tracks as training assistants in subsequent training. At the same time, six architects and engineers managed to secure employment either with UNESCO, ICCROM or with contractors engaged in rehabilitation projects. Further evidence of the impact can be observed through post-training assessments among craftspeople: four out of five blacksmiths experienced a notable income increase, surpassing a twofold increment in some cases. Comparable patterns provide empirical support for the programme's impact on livelihoods.

6 Key Success Factors

The success of the course in linking heritage capacity building with livelihoods and economic recovery in post-conflict Mosul, Iraq can be attributed to several key factors. Firstly, a participatory need assessment was conducted, which involved consultations with local stakeholders to ensure that the training provided was aligned with the real needs on the ground. This approach ensured that the programme was relevant and responsive to the specific context of Mosul. Secondly, the course adopted a multi-track approach targeting professionals and craftspeople. This strategy addressed Mosul's diverse economic rehabilitation challenges, catering to different skill levels and occupational needs. The programme also employed a hands-on methodology, incorporating site visits and practical activities into the training. This approach provided authentic learning experiences and made the training relevant to the participants' work.

Inter-generational knowledge transfer was another critical factor in its success. The programme leveraged the expertise of master craftspeople to promote continuity in traditional skills, ensuring that these valuable skills were not lost. The course also provided toolkits, books and mentorship to participants, enabling them to apply their knowledge and skills after graduation. This support was crucial in ensuring that the training had a tangible and immediate impact on the participants' livelihoods. Local partnerships were also instrumental. The programme engaged with contractors, universities, and the government to facilitate participant recruitment and job placements. These partnerships ensured that the training was directly linked to employment opportunities. Finally, the course demonstrated flexibility in its training curriculum, adapting to evolving needs and employment trends in Mosul's post-conflict economy. This adaptability ensured that the programme remained relevant and effective in changing circumstances.

7   Conclusion

The preliminary evidence of the success of the ICCROM Heritage Recovery Programme, as demonstrated by the number of graduates finding employment and initiating cultural development projects, shows that investment in local capacity building can yield extensive benefits in post-conflict reconstruction. It has improved livelihoods and empowered the youth of Mosul to regain a sense of identity and belonging through their cultural heritage. The approach and impact of the programme provide valuable insights for rebuilding initiatives in other conflict-affected areas. Heritage conservation can directly support economic recovery and community resilience if local capacities are mobilised through participatory, needs-based training.

As Mosul rebuilds, the professionals and craftspeople trained through the programme will continue to make invaluable contributions, fuelling the city's revival. The most significant impact is the empowerment of Mosul's youth to reconnect with their heritage. By equipping them with knowledge and skills, the programme enabled them to actively contribute to their city's rehabilitation. It is helping lay the groundwork for the youth of Mosul to manage its heritage sustainably. The strategic focus on capacity building and livelihood creation aligns with the "building back better" approach of the "Revive Spirit of Mosul" initiative. ICCROM's training programme embodies this holistic vision of forging a resilient future for Mosul's citizens and cultural heritage. The programme serves as a model for heritage-based livelihood regeneration post-conflict. Investing in youth and artisans generates social and economic returns crucial for recovery. The local empowerment approach ensures that solutions are context-specific and sustainable.

References

ICCROM (2021a) Heritage recovery programme in Mosul. https://www.iccrom.org/courses/heritage-recovery-programme-mosul

ICCROM (2021b) Heritage recovery in Mosul. https://www.iccrom.org/news/heritage-recovery-mosul

ICCROM (2023) Graduation ceremony for craftspeople and exhibition on ICCROM's Mosul heritage recovery programme. https://www.iccrom.org/news/graduation-ceremony-craftspeople-and-exhibition-iccroms-mosul-heritage-recovery-programme

UNESCO (2023) Restoring livelihoods through culture in Yemen. https://www.unesco.org/en/articles/restoring-livelihoods-through-culture-yemen-0

UNESCO (n.d.) Expert meeting on economic dimensions. https://ich.unesco.org/en/livelihoods-01315

UNESCO in Afghanistan: Historical Highlights from Monument Conservation to Debates on Reconstruction

Nao Hayashi

Abstract UNESCO has been collaborating with the government of Afghanistan and partners such as academia and civil society since the 1950s, in the fields of built heritage, World Heritage, museums, intangible and living heritage. Since 2001, it has focused on the safeguarding of two World Heritage sites in Afghanistan, Bamiyan and Jam, also placed on the List of World Heritage in Danger. Building on its previous achievement thanks to the cooperation with various partners, the international community continues to provide various supports, including to fulfill the conditions for removing the two sites from the List of World Heritage in Danger, as well as to further discuss the issue of reconstruction, as Bamiyan represents a unique case for reflection on this subject.

Keywords Afghanistan · World heritage · Heritage and conflict

The ideas and opinions expressed in this article are those of the author and do not necessarily represent the view of UNESCO.

1 Early Dates

The purpose of this short article is to provide a brief historical overview of the actions of UNESCO in Afghanistan, from its early days to the present. It is a story marked by the vagaries of global geopolitics. But it is also woven, despite the challenges of the circumstances, by a remarkable will of actors around the world, in particular experts, academia, and civil society, to preserve this nation's heritage and history.

UNESCO's intervention in Afghanistan for its heritage dates to the early 1960s, at the request of the Afghan government to save its heritage. Among the very first projects, was the cooperation for the Minaret of Jam, "rediscovered" by the French

N. Hayashi (✉)
UNESCO World Heritage Convention Programme Specialist, Paris, France
e-mail: n.hayashi@unesco.org

© The Author(s) 2024
M. Loda and P. Abenante (eds.), *Cultural Heritage and Development in Fragile Contexts*, Research for Development, https://doi.org/10.1007/978-3-031-54816-1_18

Archaeological Delegation (DAFA) team in 1957. Missions to the Jam site in 1962 and 1963 carried out emergency conservation work with the help of local villagers. This masterpiece of Islamic heritage was declared a World Heritage Site in 2002[1] and celebrates the 20th anniversary of its inscription in 2022.

It was also the time of UNESCO's first and largest campaign—that of saving the monuments of Nubia at Abu Simbel, whose great catalyst was a French minister and novelist André Malraux. Afghanistan, like many other countries, was inspired by this monumental campaign and aspired to a similar global action.

Thus, the campaign for the Old City of Herat began in 1976, with the approval of the General Conference of UNESCO. This project, besides the on-site training programme, included the restoration of the ancient citadel of Herat, Qala-e Ikht-yaruddin and the *madrasa* of Sultan Husain Baiqara, as well as a programme to establish an inventory of the monuments.

However, the civil conflict in Herat led by Ismail Khan in March 1979 against the communist government, as well as the outbreak of the war in Afghanistan, interrupted the work.

In the 1970s, conservation work continued at the Jam site. The UNESCO mission in 1974 carried out the first drilling around the Minaret of Jam, in order to develop a restoration plan. As work progressed to reinforce the banks and foundations, war broke out and it would be 20 years before UNESCO could resume its work on the site.

In 1989, after the withdrawal of Soviet troops from Afghanistan, the international community faced new challenges in rebuilding a country devastated by a decade of civil strife and war. In the field of cultural heritage, UNESCO immediately sent assessment missions to Balkh, Bamiyan, Herat and Mazar-I Sharif.

In a message sent in April 1990 to Federico Mayor, then Director-General of UNESCO, Prince Sadruddin Aga Khan, United Nations Coordinator for Afghanistan, appealed that despite the pressing need for relief commodities and rehabilitation projects in Afghanistan, he felt that the United Nations should never lose sight of the country's unique cultural heritage.

To rebuild a stronger and more resilient society, the culture of the past should be preserved and appreciated, culture being the memory of a people and that of a world. To this end, in parallel with the programmes for the rehabilitation of monuments and sites, the revitalization of the country's traditional crafts has been launched. One example is a training workshop on traditional Afghan carpets held in Mazar-I Sharif in 1990, in cooperation with the Afghan Carpet Exporters Guild.

Thus, little by little, we advance towards our time.

After the fall of Najibullah's government in 1992 and the creation of the Islamic State, through the Taliban seizure of power from 1996 until 2001, UNESCO continued to deploy its efforts mainly through non-governmental organizations.

---

[1] https://whc.unesco.org/en/list/211/, last accessed 2023/09/22.

2 The Destruction of the Buddhas, a Historic Turning Point and Iconoclastic Waves Since 2001

The gravity of the March 2001 event left a deep impression on our minds. The destruction of the two Buddhas was distinguished from the destructive acts of monuments and the spoliation of art objects practised since Antiquity: this time it was an act put on display and a desire to suppress the testimonies of exchanges and interconnection between different cultures and beliefs.

At the time, UNESCO and its Director-General Koïchiro Matsuura mobilized the leaders of the Islamic world, who responded to this call by deploying deterrence campaigns. Unfortunately, these collective efforts have not stopped the Taliban from continuing to try to destroy the region's heritage.

This iconoclastic destruction has followed one another in waves since this event, in Syria, Iraq, then in Libya, Mali and elsewhere. Paradoxically, they testify to the importance of heritage, since their existence represents an evocative power of memory, of our common and shared past. UNESCO and the Government of Afghanistan established an International Coordinating Committee for Afghanistan in 2003, and the site of Bamiyan was inscribed on the World Heritage List in 2003, following the inscription of the Minaret and Archaeological Remains of Jam in 2002.

3 Bamiyan and Jam

Since the destruction of the Bamiyan Buddhas in 2001 until today, UNESCO has played the role of coordinating the efforts of the international community to rehabilitate the cultural property in Afghanistan, in the different provinces of the country. More than US $27 million has been invested, inter alia, for the conservation and stabilization of the site, the empowerment of local communities, the revitalization of intangible cultural heritage, and the enhancement of creativity.

3.1 Restoration of the Buddha Niches and Component Sites of Bamiyan

For Bamiyan, UNESCO led six successive phases of a project, from 2001 to the present, funded by the government of Japan, for the stabilization of the niches where the Buddhas were located, which were in danger of collapsing. After more than 15 years, the work to consolidate the eastern niche was completed, while very urgent work advanced to save the western niche, until August 2021.

Before 2021, the team installed ten rows of steel ropes in the top part of the former head-neck area and the industrial climbers, hanging on a rope from the top of the former head area of the Western Buddha statue, managed to drill holes for the anchor points by leaning on the existing protecting steel net. The temporary measures will allow for the extension of the scaffolding up to the top of the niche, and for upcoming operations of grouting, nailing, and anchoring of the loose rock materials.

A series of workshops were led by government officials and UNESCO experts in four different community areas in Bamiyan (Qala e Ghamay, Tolwara, Sukhqohl, and Jogra Khel) in order to involve around 500 people from local communities in the implementation of the Cultural Master Plan.

The issue of buffer zone management and regulating urban expansion in the Valley remains one of the challenges, to preserve the site as a cultural landscape by possibly bringing to the modification of the boundaries in order to include heritage assets and resources which are not included in the current zoning.

The other project, supported by Italy, focuses on the safeguarding of the fortress of Shahr-i-Gholghola, a component site of Bamiyan, including research and training for caves covered with murals, remarkable expressions of Indian and Chinese influences encountered on the Silk Road.

Technical working meetings on Bamiyan were held 14 times between 2003 and 2018, and an international conference in 2017 in Tokyo was dedicated to the question of a possible reconstruction of the destroyed statues.

3.2   Preservation of Minaret of Jam

For Jam, "Emergency consolidation and Restoration of Monuments in Herat and Jam, Phase I and II" were implemented from 2003 to 2012, under the Italian funding. The project aims at the emergency consolidation, conservation and restoration of the Minaret of Jam and The Fifth Minaret in Herat. The long-lasting cooperation achieved, among others, the consolidation and rehabilitation of Minaret of Jam and Fifth Minaret in Herat, increased national capacity in the conservation of cultural heritage, income generation and training provided for Afghans of various back-grounds, in particular craftsmen and workers, thus improving the living conditions of local communities.

The Expert Working Group for the Old City of Herat and World Heritage property of Jam was established in 2002, with the mandate of coordinating all cultural projects in the country entrusted to UNESCO by the Afghan government. The Group co-ordinates activities carried out under the funds provided by Italy, Swiss and Norway to UNESCO, and reviewed other bilateral activities carried out in Afghanistan by other agencies such as the Aga Khan Trust for Culture over the previous years.

In September 2012, at the end of the 3rd Expert Working Group Meeting held at the Museo d' Arte Orientale in Turin, Italy, the project's final recommendations were drawn for future safeguarding actions for the two sites by more than 30 experts and high-ranking officials from Afghanistan, as well as representatives of donor countries.

Such a mechanism of Expert Working Group, like the one on Bamiyan, greatly enhanced synergies between the initiatives underway in Afghanistan to help safeguard the country's cultural heritage.

The history of international cultural cooperation is in detail recounted in two books: the first, *Safeguarding the Cultural Heritage of Afghanistan: Jam and Herat*, was published by UNESCO in 2015[2] and a more recent book entitled *The Future of the Bamiyan Buddha Statues Heritage Reconstruction in Theory and Practice* co-published by Springer and UNESCO in 2021,[3] collecting high-quality scientific articles on the extremely complex issue of reconstruction.

3.3   The Case of Bamiyan: Reflection on the Reconstruction and Reinterpretation of an Archaeological Site

The two giant Buddha statues and associated archaeological features in Bamiyan had become an iconic and integral part of the wider cultural landscape in the Bamiyan Valley, and long after the decline of Buddhism in the region, they had continued to be a source of national pride and integrated into local traditions and folklore. Their destruction was a significant loss to the country and to humankind.

Inscribed on the List of World Heritage in 2003, the "Cultural Landscape and Archaeological Remains of the Bamiyan Valley" was simultaneously placed on the List of World Heritage in Danger. As a serial World Heritage property, it consists of eight separate component sites located in the Valley and its tributaries. Once an important strategic location between East and West on the Silk Roads, the property also includes many caves that form several ensembles of Buddhist monasteries, chapels and sanctuaries dating from the third century to the thirteenth century C.E., representing the artistic and religious developments of ancient Bactria, and integrating various cultural influences into the Gandharan school of Buddhist art. The Buddhist monastic ensembles and sanctuaries, as well as fortified structures from the Islamic period, testify to the interchange of Indian, Hellenistic, Roman, Sasanian and Islamic influences.

A significant proportion of the attributes that express the property's Outstanding Universal Value remain intact after the attacks, including the vast Buddhist monastery in the Bamiyan Cliffs which contained the two colossal sculptures of the Buddha.

Destruction and Outstanding Universal Value (OUV)

In the framework of a discussion on reconstruction, it is worth noting that the World Heritage Committee inscribed the Bamiyan Valley under criterion (vi), "directly or tangibly associated with events or living traditions, with ideas, or with beliefs, with artistic and literary works of outstanding universal significance". The Statement of OUV notably highlights that "Due to their symbolic values, the monuments have

---

[2] https://unesdoc.unesco.org/ark:/48223/pf0000233042, last accessed 2023/09/22.

[3] https://unesdoc.unesco.org/ark:/48223/pf0000375108.locale=en, last accessed 2023/09/22.

suffered at different times of their existence, including the deliberate destruction in 2001, which shook the whole world". Therefore, the attempts to destroy parts of the property at various points in time, including by the Taliban, form an integral part of the OUV for which the property was inscribed. This element has a profound bearing on any discussion about a possible reconstruction of the Bamiyan Buddha statues, as the property was recognized as being of Outstanding Universal Value after and for its partial destruction.

During the 40th session of the World Heritage Committee in Istanbul (Turkey) in 2016, the Government of Afghanistan requested that at least one of the Buddha statues be reconstructed. However, heritage practitioners and experts have highlighted the challenges of reconstructing the Buddha statues according to strict conservation ethics and other considerations. At the Committee's 41st session (Krakow 2017), general decisions concerning reconstruction (41 COM 7) were also taken.

Responding to the request of the Government of Afghanistan and the decisions of the World Heritage Committee concerning the deliberate destruction and reconstruction of cultural heritage in areas of conflict, in particular the Buddha statues in Bamiyan, UNESCO, together with the Government of Afghanistan and the Tokyo University of the Arts and with a financial support of the Government of Japan, organized an international technical meeting entitled "The Future of the Bamiyan Buddha Statues: Technical Considerations & Potential Effects on Authenticity and Outstanding Universal Value" in Tokyo, Japan from 27 to 30 September 2017. These meetings aimed to discuss the potential reconstruction of the Buddha statues in the Bamiyan Valley and provided an ideal forum to discuss and clarify the current theory and practice surrounding the reconstruction of cultural property and restoration ethics; they also allowed to address the issue of authenticity and the impact on the OUV of the Bamiyan World Heritage property by such restorations and reconstructions. Through a series of presentations delivered by international experts, the Forum concluded that further consultation and reflection are needed to address relevance and feasibility of the reconstruction.

Warsaw International Conference on Reconstruction: The Challenges of World Heritage Recovery in 2018

A conference entitled "The Challenges of World Heritage Recovery. International Conference on Reconstruction" was organized in Warsaw, Poland, in May 2018. The participants adopted the Warsaw Recommendation on Recovery and Reconstruction of Cultural Heritage, which notably highlights that "Memorialization of the destruction should be considered for communities and stakeholders; … In the context of post-conflict recovery and reconstruction, such places should integrate as much as possible a shared narrative of the traumatic events that led to the destruction, reflecting the views of all components of the society, so as to foster mutual recognition and social cohesion, and establish conditions for reconciliation". The Declaration also points out that "The key to a successful reconstruction of cultural heritage is the establishment of strong governance that allows for a fully participatory process, is based on a comprehensive analysis of the context and on a clear operational strategy, including mechanisms for the coordination of national and international actors, and is supported by an effective public communication policy".

Reconstruction of cultural properties in the context of the World Heritage Convention

Reconstruction of cultural properties is outlined in the Operational Guidelines of the Convention as follows: "... reconstruction of archaeological remains or historic buildings or districts is justifiable only in exceptional circumstances. Reconstruction is acceptable only on the basis of complete and detailed documentation and to no extent on conjecture" (UNESCO World Heritage Center 2023). Given the current move towards heritage reconstruction in post-conflict contexts, some more detailed guidance is required to reflect the multi-faceted challenges that heritage reconstruction brings, especially with regard to the OUV of the World Heritage properties.

Any further exploration of reconstruction should be made in very close consultation with impacted communities who have particular connections with heritage and suffered from its loss, in order to understand the meaning of this heritage for them, incorporate the multiplicity of interpretations of heritage, and determine whether they wish to rebuild, reconstruct and re-establish such heritage properties.

If there is community consensus for reconstruction, a number of additional inquiries arise, such as how it will be done; who will make decisions; and what purpose the reconstruction may fulfil?

Moreover, when it comes to the inscribed World Heritage properties, how can reconstructed heritage using new materials be considered to retain authenticity? Reconstruction of cultural heritage requires not only in-depth discussion in a multi-disciplinary approach by experts and academia but also a long-term vision for the preservation and interpretation of such reconstructed heritage, which duly requires the involvement of communities who have suffered from damaged and/or lost heritage.

Specific concerns for Bamiyan

The inscription of Bamiyan took place after the destruction of Buddhas. This particular situation prompts us to ask: what to reconstruct, what to preserve?

Additionally, the reconstruction of the Great Buddha is discussed in the context of Muslim country. It should be noted that the two Buddha statues of Bamiyan were not considered as religious figures by the inhabitants of the Bamiyan Valley for centuries, but instead as a prince of Bamiyan and a princess of foreign extraction. As such, the two statues have dominated the landscape of Bamiyan for centuries and left their mark on local legends and folklore. We understood that the desire of inhabitants of the Bamiyan Valley to see the rehabilitated Buddha statues has been motivated by this and not by the desire to reconstruct statues in the perspective of religious worship.

In addition, while raising the issue of intentionally destroyed cultural heritage, there is a huge difference between the reconstruction of a Buddha statue in human form (idol), considered as alive, and the reconstruction of a building that is recognized as having no life in the first place.

4   Today: Actions to Remove Bamiyan and Jam from the List of World Heritage in Danger

Today, UNESCO is paying particular attention to guiding actions in Bamiyan and Jam for the removal of properties from the List of World Heritage in Danger. This requires fulfilling a number of the major conditions adopted by the World Heritage Committee: for Bamiyan, among others,

- ensured site security;
- ensured long-term stability of the Giant Buddha niches;
- adequate state of conservation of archaeological remains and mural paintings; and
- implemented Management Plan and Cultural Master Plan (the protective zoning plan).

For the Jam site,

- Increased capacity of the staff of the Afghan Ministry of Culture and Information in charge of the preservation of the property ensured;
- Precisely identified World Heritage property and clearly marked boundaries and buffer zones;
- Long-term stability and conservation of the Minaret of Jam ensured;
- Site security ensured;
- A comprehensive management system including a long-term conservation policy developed and implemented.

The ongoing project at Jam, funded by ALIPH, is aimed at developing a comprehensive conservation plan focusing on its overall protection and monitoring, including the design and implementation of emergency actions to strengthen the structural stability of the Minaret.

5   Museums and Collections

Another area of UNESCO's intervention in Afghanistan concerns the museum sector. Readers probably remember that famous slogan hanging at the entrance of the National Museum of Afghanistan, in the aftermath of the fall of the Taliban regime in 2001, saying "a nation stays alive when its culture stays alive".

Since its inception in the 1930s, the National Museum in Kabul has shone as a place of exchange on an international scale.

Since 2001, many organizations and donors from the cultural sector have been working with UNESCO to restore the heavily damaged building of the National Museum and its collections: DAFA, Musée Guimet in Paris, Austrian Society for Afghanistan, British Museum, National Geographic Society, International Security Assistance Force (ISAF), Hellenic Aid, International Council of Museums (ICOM) as well as the governments of Italy, Japan and the United States.

In addition, in order to combat the scourge of illicit trafficking of works of art, UNESCO and its partners have focused their efforts on the rehabilitation of collections and the compilation of documentation from inventories of the National Museum. "Catalogue of the National Museum of Afghanistan, 1931–1985"[4] was published in 2006 to save the records on national collections in case of their loss, theft and in preparing for their possible return.

Some will remember that the Museum-in-Exile in Switzerland or the Hirayama Foundation in Japan housed valuable art objects evacuated from Afghanistan in their respective institutions during these turbulent times and returned these collections to the national authorities. The first was a large collection of ethnographic objects, and the latter included fragments of archaeological pieces from various sites, including Ai Khanum, a Hellenistic city said to be built by the Seleucid successors of Alexander the Great.

In Gazni, eastern Afghanistan, the Museum of Islamic Art and the Museum of Pre-Islamic Art have been rehabilitated, helping to house monuments and art objects.

This Italian-funded project through UNESCO contributes to the inventory and enhancement of collections from the Buddhist site of Tapa Sardar, dating from the Kushan period, thanks to the technical support of ISIAO—Institute of Africa and the Orient and the Afghan Department of Historical Monuments.

The Cultural Centre for Bamiyan, whose construction was recently completed in 2022, is being built with Korean funding. Intended to safeguard artefacts, promote access to intangible culture and knowledge on heritage, its programming and concrete activities remain to be determined soon.

Due to the security situation in Afghanistan, which remains highly volatile and has been further exacerbated by the COVID-19 pandemic and the de facto government takeover since August 2021, field operations require a number of adjustments. The interim Transitional Engagement Framework (TEF) recently terminated and the Strategic Framework for Afghanistan (UNSAF) was endorsed in July 2023 for a period of two years. This Strategic Framework articulates the UN's approach to addressing basic human needs of the people of Afghanistan.

UNESCO, through its office in Kabul, continues to monitor the situation on the ground and seizes all possible opportunities to sensitize political and administrative actors on the importance of heritage.

Throughout UNESCO's institutional history and in dialogues with its Member States, the challenge of mutual understanding lies at the confluence of the orders of cultural values and relations between peoples.

Afghanistan was and remains today as a crossroads of cultures, beliefs, and trade over millennia. Its art and heritage bear witness to this incessant intermingling between Europe, Asia and the Middle East.

Our goal remains, not only to restore the heritage, the finest treasures of art and archaeology of Afghanistan and Central Asia, but also to allow the culture of the country, with its diversity and richness, to regain its pre-eminent place in Afghan

---

[4] https://unesdoc.unesco.org/ark:/48223/pf0000148244, last accessed 2023/09/22.

society as an essence testifying to the complexity of multicultural identity, past and present, of this nation.

References

Decision 41 COM 7 of the World Heritage Committee, https://whc.unesco.org/en/decisions/6940/
Grissman C (2002) The inventory of the Kabul Museum: attempts at restoring order. Mus Int LV(3–4):71–76
UNESCO (2015) Safeguarding the cultural heritage of Afghanistan: Jam and Herat
UNESCO (2021) The future of the Bamiyan Buddha statues heritage reconstruction in theory and practice. Springer and UNESCO
UNESCO World Heritage Center (2023) Operational Guidelines to the World Heritage Convention, p. 31
Warsaw Recommendation: https://whc.unesco.org/en/news/1826/

ALIPH in Iraq

Elsa Urtizverea

Abstract Iraq is the *raison d'être* of ALIPH, and since the organization became operational in Geneva in September 2018, the foundation has supported more than 40 projects in Iraq. Combined, these projects have the goal to rehabilitate monuments and sites, safeguard museums and their collections, preserve and enhance written heritage, protect intangible heritage, and combat looting and illicit trafficking of cultural property. ALIPH is able to carry out its mission to help protect Iraqi heritage thanks to its close working relationships with the Iraqi Ministry of Culture, Tourism and Antiquities and the State Board of Antiquities and Heritage, with whom the Foundation shares a common holistic approach toward safeguarding and conserving heritage from its roots, stretching back thousands of years, to its future.

Keywords ALIPH · Rehabilitation · Peace building

1 Cradle of Civilization

Iraq stands as a living testament to the fascinating intersections of cultures, beliefs, and traditions that have shaped the region's identity over thousands of years. With a history spanning ancient Mesopotamia, the rise and fall of great empires, and the emergence of Islam, Iraqi cultural heritage is a mosaic of influences from the Sumerians, Babylonians, Assyrians, Persians, and Arabs, and countless others.

Iraq is widely regarded as the "Cradle of Civilization" due to its association with the first human settlements and the development of advanced civilizations. The fertile lands between the Tigris and Euphrates rivers provided the foundation for the Sumerian and Babylonian civilizations, notably, the birthplace of the written form.

Iraq is also home to myriad religious traditions, making it a significant hub of spiritual diversity. Islam is the predominant religion, with both Sunni and Shia communities represented and numerous religious sites, but Iraq is also home to ancient religious

E. Urtizverea (✉)
International Alliance for the Protection of Heritage in Conflict Areas, Geneva, Switzerland
e-mail: elsa.urtizverea@aliph-foundation.org

© The Author(s) 2024
M. Loda and P. Abenante (eds.), *Cultural Heritage and Development in Fragile Contexts*, Research for Development, https://doi.org/10.1007/978-3-031-54816-1_19

Fig. 1 Arch of Ctesiphon—December 2021—© ALIPH—Azhar al-Rubaie

communities of Christians, Yazidis, Mandaeans, and others, each contributing to the rich tapestry of religious practices and beliefs that form a part of Iraqi patrimony.

Cultural heritage is a fundamental element of Iraqi identity. It shapes the sense of belonging, instilling shared historical narratives that transcend individual differences. Thanks to this richness, Iraq can boast architectural marvels that showcase the ingenuity of its ancient civilizations. The Ziggurat of Borsippa, the Mashki Gate (Nineveh), the monumental Arch of Ctesiphon (Taq Kasra), and the ruins of Hatra, to name but a few examples, are enduring symbols of the grandeur and sophistication of Iraq's architectural heritage. Islamic architecture further enriches the country's cultural landscape, reinforcing the connection between modern Iraqis and their ancestors and fostering a deep-rooted sense of heritage and pride (Fig. 1).

2   Iraq, the *Raison D'être* of ALIPH

While the diversity and richness of Iraqi culture are at the heart of its appeal and fascination throughout the world, they have also, sadly, fallen victim to devastation and destruction during the recent conflicts that have affected the country. Over the last few decades, this priceless heritage has suffered immensely from terrorism, war, and instability.

The devastating attacks perpetrated by Daesh against heritage in Iraq have had profound consequences, not just for the country but for the entire world. These acts of deliberate destruction, sadly echoing similar acts perpetrated in Mali, have marked a turning point of profound and multidimensional significance—heritage has transformed from a collateral victim in conflict to an effective weapon of war.

This destruction is an intentional attempt to erase culture, and by doing so, eradicate the essential elements of a country's historical identity. Further, it is also cynically used as a propaganda weapon, through the broadcasting of videos designed to illustrate the violence, power, and determination of terrorist groups. In addition, these acts of destruction can serve as a source of funding: the looting and illicit trafficking of artifacts has served terrorist groups as a financial resource and can strengthen their position and influence.

This massive destruction of heritage in the Sahel and the Middle East has left the world in turmoil. Given the scale of the task and the need for an emergency mechanism, the United Arab Emirates and France took initiative and decided to create a new foundation: the International Alliance for the Protection of Heritage in Conflict Areas, ALIPH.

Iraq is therefore the *raison d'être* of ALIPH, and since the organization became operational in Geneva in September 2018, the foundation has supported more than 40 projects in Iraq. Combined, these projects have the goal to rehabilitate monuments and sites, safeguard museums and their collections, preserve and enhance written heritage, protect intangible heritage, and combat looting and illicit trafficking of cultural property.

3   A Tapestry of Projects

ALIPH is able to carry out its mission to help protect Iraqi heritage thanks to its close working relationships with the Iraqi Ministry of Culture, Tourism and Antiquities and the State Board of Antiquities and Heritage, with whom the Foundation shares a common holistic approach toward safeguarding and conserving heritage from its roots, stretching back thousands of years, to its future. Over the past 5 years, ALIPH has also established numerous partnerships with local and international operators and prioritizes the inclusion of local actors and communities in the development and implementation of all projects.

Heritage protection is a lever for job creation, tourism, and sustainable economic and social development. For ALIPH, protecting Iraq's exceptional heritage also means contributing to on-the-job training, especially for young people, in the fields of construction, crafts, archaeology, and conservation. In parallel, this work is a way to contribute, whenever possible, to dialogue and reconciliation among communities torn apart by conflict.

From the conservation of the site of Hatra, the stabilization of the Arch of Ctesiphon, and the restoration of the Khan Marjan in Baghdad, all of which bear witness to the rich cultural and architectural history of Iraq, to the preservation of Iraq's priceless written heritage, ALIPH has supported over 40 projects to protect or rehabilitate Iraqi cultural heritage since 2018. A dozen of these projects have been completed, over 20 are in the Nineveh Governorate, and 11 are in the city of Mosul.

4   Mosul Mosaic and the Mosul Museum

The city of Mosul, which means "the junction" in Arabic, has been characterized by a multi-religious and multicultural environment for centuries. From 2014 to 2017, Mosul fell prey to violent extremism, leaving the city and its urban heritage in ruins.

In February 2018, UNESCO launched the "Revive the Spirit of Mosul" program. To contribute to this initiative, ALIPH established its "Mosul Mosaic" strategy to help preserve cultural heritage of Mosul in all its diversity. Five landmarks that have made up Mosul's skyline for centuries are being rehabilitated under this initiative: the Al-Raabiya Mosque (operators: Archi.Media Trust Onlus, SBAH, Mosul University, Waqf authorities, University of Florence), the Al Musfa Mosque (operators: La Guilde Européenne du Raid, SBAH, Waqf authorities, and the Institut National du Patrimoine (INP)), the Mar Toma Syriac Orthodox Church (operators: Œuvre d'Orient, SBAH, the Syriac Archbishop of Mosul, INP), the Chaldean Church Al Tahira (operators: Œuvre d'Orient, SBAH, the Chaldean Archbishop of Mosul, INP), as well as the remarkable Tutunji House (operators: University of Pennsylvania, SBAH, Mosul University) (Figs. 2 and 3).

In addition to these projects, and chief among Mosul's rich and diverse cultural heritage, is the rehabilitation of the Mosul Cultural Museum (Fig. 4).

Fig. 2   Mar Toma Church © G. de Beaurepaire et G. de Salins, L'Oeuvre d'Orient

Fig. 3 Tutunji House © Elsa Urtizverea—ALIPH

Fig. 4 View of the main façade of Mosul Cultural Museum from the cleared garden (2018) © SBAH

The Mosul Museum is one of the largest museums in Iraq, second only to the National Museum in Baghdad. Founded in 1952 and originally housed in the former palace of King Ghazi, it moved to its current building in 1974. In 2003, to protect the collection from potential looting, about 1,500 pieces were moved to Baghdad, where they remain until the reopening of the museum. Only monumental pieces stayed in the Mosul Museum. The Museum was attacked by looters, but the army quickly regained the control. In 2014 the Museum was occupied by Daesh. In February 2015, the extremists publicized videos showing the destruction of its artifacts. The Museum was liberated by the Iraqi army 2 years later (Fig. 5).

In 2018, at the request of the Iraqi authorities, ALIPH initiated the rehabilitation of the Museum. This ambitious initiative—one of ALIPH most important project to date—is carried out in partnership with the Ministry of Culture, Tourism and Antiquities of Iraq, the State Board for Antiquities and Heritage and the Mosul Museum. The Louvre Museum is responsible for restoring the collection and creating new museography, and the Smithsonian Institution conducted the first technical assessment of the situation and is leading the capacity building for the Museum staff. In the first stage of the project (2018–19), immediate stabilization measures were implemented, including removing unexploded ordnance from the roof, shoring up collapsing floors, and securing the building. The objects and fragments in the collection were sorted, cleaned, documented, and stored. In 2020, despite the pandemic, the conservation of the artifacts progressed. The Musée du Louvre prepared and carried out remote

Fig. 5 The Mosul Museum © Elsa Urtizverea—ALIPH

training in conservation techniques for the museum staff. In 2021, World Monuments Fund joined the consortium as a partner responsible for the rehabilitation of the Museum's building. ALIPH finances each component of this project and provides significant scientific and technical oversight across all activities. Closed since 2003, the project aims to have the Museum reopen in 2026.

The building, an important piece of Iraqi cultural heritage in its own right, is today a symbol of life overcoming ruin. It is a physical reminder of how peace and sustainable development can prevail over acts of violence and inhumanity.

The project to rehabilitate this building and its collections is extraordinary in its scope, in the sheer number of technical challenges that are being overcome, and in how people of different nationalities, languages, and identities have come together to achieve a common mission.

This museum was once home to a collection of artifacts that showcased the long and rich history of the Northern Iraq Region and—more broadly—of all of Mesopotamia. Before our very eyes, this extraordinary collection was savagely ransacked by Daesh. While the purpose of that targeted destruction might have been to erase millennia of history, these artifacts are now proudly rising from the ashes, almost literally, thanks to the exceptional work and expertise of the restorers and the commitment of the Mosul Museum team. Those images of destruction, as tragic as they were violent, are gradually giving way, fragment by fragment, to a vision of a brighter future.

If Daesh thought they had accomplished their mission, then the mobilization of this multidisciplinary teams from Iraq and abroad, is proving them wrong. Not only will this museum once again tell the multi-millennial history of the region, but it will also bear the scars of a recent war in which heritage was counted among the victims and will live on as a testament to humanity's will to overcome tragedy (Fig. 6).

"As representatives of the Heritage Department, we are happy to see that the Louvre Museum, the Smithsonian Institution, the United Arab Emirates, France, and other countries are willing to contribute to the reconstruction of what was destroyed by the militia. We have a sense of pride in this land, which is important not just for the Iraqi people, or us—the employees, but also for the international community. It is everyone's responsibility to save this heritage and this cultural identity," said Zaid Ghazi Saadallah the Director of the Museum during a press conference in May 2023 to announce the beginning of the building's rehabilitation and its reopening in 2026.

5   Sites and Discoveries

The former caravan city of Hatra (northern Iraq, 100 km south of Mosul), a UNESCO World Heritage Site, reached its pinnacle between the second and third century CE, as a religious center for the cult of Shamash, the God of the Sun. For centuries it was the focus of bitter conflicts between Rome, the declining empire of the Parthians, and the rising Sasanid Empire. At the beginning of 2015, the archaeological area came under Daesh occupation when it was used for military training, and its artifacts were

Fig. 6 The Mosul Museum © Elsa Urtizverea—ALIPH

targeted by jihadist militants. The most irreplaceable and unique Hatrene art was
vandalized and severely damaged, and its destruction broadcast around the world by
Daesh: acts declared a "war crime" by UNESCO. The site was added to the List of
World Heritage in Danger on 1 July 2015.

Since 2017, ALIPH has been supporting an Italian-Iraqi scientific team that was
the first to enter the site after the occupation by Daesh. The mission—composed of
archaeologists and architects led by Prof. M. Vidale (University of Padua) and Prof.
S. Campana (University of Siena), working under the aegis of ISMEO (International
Association for Mediterranean and Eastern Studies—Rome), in partnership with
the State Board of Antiquities and Heritage of Iraq, the University of Padua and

the University of Siena—has now thoroughly documented and mapped the site of over 700 hectares using drone imagery. In addition, fragments of the vandalized sculptures have been recovered and secured. The most spectacular discovery was fragments from large sculpted heads that had decorated the facade of the sanctuary dedicated to the God of the Sun. Their brutal destruction was widely disseminated on social media by Daesh. The larger fragments of the sculptures fit perfectly, making it possible to restore and relocate them back to their original position (Fig. 7).

In addition, ALIPH supported two projects in Mosul and the Nineveh region that have led to significant archaeological discoveries. A project implemented by the University of Pennsylvania, in cooperation with the State Board for Antiquities and Heritage and Mosul University, to reconstruct the exceptional Mashki Gate in Mosul—one of the monumental gates of Nineveh—uncovered eight reliefs from the era of the Assyrian King Sinharib (705 to 681 BCE) in late 2022.

In Faida, north Kurdistan, a project conducted by the University of Udine together with the Directorate of Antiquities of Duhok, led to the discovery of ten monumental rock reliefs dating from the eighth-century BCE and depicting the ruler and great gods of Assyria. This discovery was awarded the prestigious 2020 International Archaeological Discovery Award "Khaled al-Asaad." Other large-scale projects have been implemented with ALIPH's support in the Nineveh Governorate, such as the restoration of the Mar Behnam monastery and several Yezidi temples in Sinjar.

Fig. 7 Hatra—ALIPH visit—May 2023 © Vincent Boisot

6 Conclusion

Even if the Foundation counts, to date, over 400 projects in 35 countries on four continents, ALIPH's engagement in Iraq remains a priority.

By swiftly responding to emergencies, funding rehabilitation projects, protecting manuscripts, intangible heritage, and putting together all the scientific knowledge of national and international experts, ALIPH has the ambition and the strong conviction that it plays a critical role in peacebuilding and economic and sustainable development. However, the journey is far from over, and continued support, cooperation, and long-term sustainable strategies are essential to secure Iraq's cultural legacy for future generations. ALIPH's commitment stands as a testament to the power of collaboration and partnership in protecting the shared heritage of humanity in the face of adversity.

7 About ALIPH

The International Alliance for the Protection of Heritage in Conflict Areas (ALIPH) is the main global fund exclusively dedicated to the protection and rehabilitation of cultural heritage in conflict zones and post-conflict situations. It was created in 2017 in response to the massive destruction of cultural heritage over the past two decades due to terrorism and conflict, predominantly in the Middle East and the Sahel. ALIPH is a public–private partnership assembling various countries and private donors. Based in Geneva, this Swiss foundation also benefits from the privileges and immunities of an international organization thanks to a headquarters agreement signed with the Swiss Confederation. To date, ALIPH has supported about 430 projects in 35 countries on four continents. ALIPH finances concrete projects carried on the ground, hand-in-hand with local partners, authorities, and communities. Its mission places cultural heritage protection as a central contributor to peace and sustainable development.

The International Institute of Humanitarian Law and Cultural Property Protection in Armed Conflict

Edoardo Greppi

Abstract The piece explores the historical path and rationale that led the international community to the negotiation and then adoption, in 1954, of the Hague Convention for the Protection of Cultural Property in the Event of Armed Conflict, focusing on some very specific peculiarities of this legal instrument, as well as on its original role within the wider IHL body of law. With a special emphasis on the provisions relating to the dissemination of the legal principles outlined by the Convention, the analysis thus presents the key elements of the research and training mission carried out, in more than half a century of activity, by the International Institute of Humanitarian Law in Sanremo (Italy) and abroad.

Keywords Hague convention · Training · Sanremo Institute

1 Codifying Cultural Property Protection: The Path to the Hague Convention

Since the end of the First World War, more than a few governments started to feel the need of codifying and draw some specific legal boundaries for the protection of cultural property in the event of armed conflict.[1]

E. Greppi (✉)
International Law at the University of Torino, Turin, Italy
e-mail: edoardo.greppi@unito.it

Master on Cultural Property Protection in Crisis Response, Turin, Italy

International Institute of Humanitarian Law, Sanremo, Italy

[1] For additional information on the legal framework of Cultural Property Protection in Armed Conflict see Frigo (1986); the key text in the IHL framework then is O'Keefe (2006). Furthermore, International Institute of Humanitarian Law (1986); Nahlik (1967, I); Toman (1984, p. 559); ID, The *Protection of Cultural Property in the Event of Armed Conflict*, Paris 1996; Panzera (1993); Gioia (2000, p. 71); Benvenuti and Sapienza (2007); Zagato (2007); Caracciolo and Montuoro (2018); Greppi (2020, pp. 133–166).

M. Loda and P. Abenante (eds.), *Cultural Heritage and Development in Fragile Contexts*, Research for Development, https://doi.org/10.1007/978-3-031-54816-1_20

Cases of voluntary destruction of cultural heritage during WWI, such as the episodes of Reims, Lovanio and Arras, did raise the issue of identifying not only a protection system at the international level but also one that would allow to monitor and ensure the military's compliance with the rules.

While already in 1915 Professors Vetter and Mariaud, respectively in Bern and Geneva, had launched the first proposals for the creation of a "Croix d'Or" (based on the idea of the Croix Rouge) to safeguard some specific monuments not to be exploited for military purposes by the parties in a conflict, the end of WWI marked the beginning of a transnational legal debate focused on the importance of developing common rules capable of harnessing the impact of war on historical buildings and works of art in general.[2]

Particularly important in this sense was the adoption of the Treaty on the Protection of Artistic and Scientific Institutions and Historic Monuments (also known as the "Roerich Pact") concluded in Washington in April 1935 and entered into force in August of the same year among the members of the Pan-American Union (later "Organization of American States"). For the first time, the Treaty indeed achieved the objective of legally defining the concepts of "respect" and "protection" of cultural objects as juridical obligations, in this way shaping the idea of a need for "general" protection both in times of peace and war.

Against the promising developments of the 1930s, the outbreak of WWII vividly underlined the extreme vulnerability of artworks and cultural property in the event of armed conflict. With the sole Hague Conventions (stipulated in 1899 and 1907) as the applicable multilateral legal framework impacting the conduct of hostilities (along with some specific restrictions foreseen, for example, by the Geneva Protocol of 1925 related to the prohibition of the use of poisonous and other gases), the total lack of international provisions regulating the treatment and safeguarding of cultural property in the event of armed conflict gave rise to a series of accidents and voluntary heinous acts (not yet formally defined as crimes).

Mostly dismissed as justified by the prevailing weight of the "military necessity" logic (for instance in the case of the destruction of the Montecassino Abbey in Italy, the bombing of the city of Dresden, or the one of Coventry), moreover, deliberate operations against cultural property often took the form, during WWII, of systematic acts aimed at erasing the heritage of a specific culture and/or population.

---

[2] In particular, a study conducted by the Archaeological Society of The Netherlands and dated 1919 served as a basis for the development of a discretionary provision within the Hague Rules of Air Warfare (1923), proposed by the Italian delegation for the protection of important cultural monuments. In the following years, these latter rules were functional to the drafting of a Preliminary Draft International Convention for the Protection of Historic Buildings and Works of Art in Time of War (1938).

As a result, the disrupted international community emerging from the war embraced the vision of consolidating specific rules to shield cultural property from hostile acts, in order to prevent civilians from being harmed by means of a direct attack against their individual and collective identity.

Hence, the new role assigned to the United Nations Educational, Scientific and Cultural Organization (UNESCO) turned out to be increasingly crucial, as it provided the States with an institutional framework to carry out the negotiation leading to the 1954 Hague Convention for the Protection of Cultural Property in the Event of Armed Conflict.

The Convention was integrated into the wider path of codification of the rules curbing the impact of armed conflict on civilians and civilian objects at the time, manifestly inheriting the legal philosophy of the four Geneva Conventions of 1949. The Hague Convention in this way perfectly fits into the overall body of IHL provisions discussed in Geneva a few years before, recalling them under several aspects such as its application framework, its implementation methodology (e.g. use of specific emblems to mark cultural property, protection of relevant personnel on the field) and the dissemination duties to be undertaken by the ratifying parties.

Through its 21 articles, the Convention not only set the overall legal boundaries aimed at protecting cultural property from direct damage in the event of armed conflict but even established that specific "refuges", used to shelter such property, and "centres containing monuments and other immovable cultural property" should have fallen under a "special protection" (art. 8–11), granting immunity from attacks thanks to the registration into an international register at UNESCO. Notwithstanding the importance of such a multi-layered approach, it is (and was at that time) self-evident how the Convention is heavily impacted by the overarching presence of the military necessity principle as a *vulnus* to protection, as well as by the several conditions required to insert the shelters in the register for the special protection.

It is also due to this weakness that, in 1977, the drafters of Additional Protocol I to the 1949 Geneva Conventions decided to reaffirm in Article 53 the importance of the provisions of the Hague Convention, recalling the prohibition of committing hostile acts, use or make the object of reprisals cultural objects and places of worship manifestly avoiding any mention of the military necessity principle. In this way, the Additional Protocol tried to take forward the idea of legal protection that should have overcome the military necessity principle, even if not explicitly defining a will to codify such development.

When, in 1999, the Second Protocol to the Hague Convention was finally adopted, the problem, however, showed off again as a persistent and still existing breach in the Hague body of law. Nevertheless, the 1999 Protocol was key to better define military necessity as "imperative" with reference to the general protection of cultural property, as well as to integrate the special protection already foreseen by the 1954 Convention with a structured "enhanced protection", much better related to military necessity through the concept of the absence of a "feasible alternative available".

Setting aside the broader international legal framework concerning the protection of cultural property in the event of armed conflict, further integrated by the two protocols in 1954 and 1999, it is key for the present analysis to move now beyond the scope of the mere physical protection of cultural property in conflict scenarios, focusing on the enlarged concept of the "safeguarding of cultural property" raised by the drafters in Article 3 of the Hague Convention.

This latter provision comes particularly useful, in fact, to better understand how deeply the drafters were aware of the paramount value of preparing "[...] in time of peace for the safeguarding of cultural property [...] against the foreseeable effects of an armed conflict, by taking such measures as [the Parties] consider appropriate".[3] While representing one of the most self-evident cases of legal complementarity between IHL and international human rights law within a humanitarian law convention (Greppi 2012),[4] thanks to the direct reference made by the text to peace as a logic antecedent of war (during which IHL becomes applicable), Article 3 essentially constitutes a first step towards the confirmation of the importance of preliminary training for the implementation of the Convention on the ground.

Nevertheless, in coherence with the approach drawn by the four Geneva Conventions, the relevance of capacity-building as a prevention instrument is even most immediately encompassed by Article 25, which outlines the dissemination duties of the contracting parties and, consequently, the need "to include the study" of the Convention's provisions "in their programmes of military and [...] civilian training".[5] In this sense, the Hague Convention and, more precisely, the formulation of Article 25 cannot help but recalling, in fact, the imprint traced by the many norms of the Four 1949 Geneva Conventions, precisely echoing not only their juridical substance but also their language.[6]

It is in this way too—and not exclusively, as previously mentioned, in the provisions referring to training and dissemination duties—that the contracting parties of the Hague Convention underlined the urgency to fill the critical gap existing in IHL concerning the protection of cultural property, namely not simply by adding a stand-alone regime to complement the Geneva Conventions but by embedding it into the comprehensive IHL body that was being defined in those very years.

---

[3] Convention for the Protection of Cultural Property in the Event of Armed Conflict (1954, Art. 3).

[4] Greppi (2012, p. 32 ss.), cf. Kolb and Gaggioli (2013).

[5] Convention for the Protection of Cultural Property in the Event of Armed Conflict (1954, Art. 25).

[6] See Geneva Convention (1949a, Art. 47, 1949b, Art. 48, 1949c, Art. 127, 1949d, Art. 144).

2   Training Military and Civilian Personnel in Cultural Property Protection in Armed Conflict: The Mission of the Sanremo Institute

Having briefly depicted a clearer view of the rationale that lies behind the codification process of a legal duty for governments to develop awareness-raising initiatives specifically addressed to their own operatives (being they military or civilians), particularly interesting now is to dwell for a moment on the experience of a unique non-governmental institution, the International Institute of Humanitarian Law. The Institute has indeed statutorily contributed for more than 50 years now to the promotion of all the IHL provisions and principles, and, among them, of the Cultural Property Protection ones.

The International Institute of Humanitarian Law is an independent, non-profit, humanitarian organization established in 1970 in Sanremo, Italy, by a group of distinguished lawyers, in a period in which the international community was launching a new diplomatic initiative aimed at the enhancement of international humanitarian law through the adoption of new, multilateral binding instruments.[7]

The main purpose of the Institute was identified in the promotion of international humanitarian law (or Law of Armed Conflicts), international human rights law, refugee law, migration law, and other related issues, through the organization of specific training programs, round tables, thematic seminars, and international gatherings addressed to national and international practitioners, military officers, diplomats, and academics.

During more than half a century of activity, the Sanremo Institute has built a global reputation as a center of excellence in the field of training, research and dissemination of IHL and IHRL, particularly thanks to the organization of hundreds of initiatives that involved more than 30,000 participants.

Over the decades, such a commitment contributed to what is nowadays recognized worldwide as the "humanitarian dialogue in the spirit of Sanremo", as well as the establishment of close collaborations with the most important international organizations dedicated to the humanitarian cause, such as the International Committee of the Red Cross (ICRC), the United Nations High Commissioner for Refugees (UNHCR), the Office of the United Nations High Commissioner for Human Rights (OHCHR), the International Organization for Migration (IOM), UNESCO, and many others.

It was exactly on the basis of these multilevel partnerships, involving regional, national, and international organizations, that the Institute has been able to support and foster the increasing international attention on the crucial issue of the Protection of Cultural Property in armed conflicts.

Since 1976 the Sanremo Institute indeed offers training programs specifically tailored to military personnel, aimed at fostering the development of knowledge in IHL. Within such courses, the Institute has always devoted special attention to

---

[7] Consider, among others, the adoption of the *Protocols Additional to the Geneva Conventions of 12 August 1949* (Protocol I and II) of 8 June 1977.

Cultural Property Protection. The topic is indeed regularly included in its annual training program, both integrated into its foundation courses and specialized training activities such as advanced courses and thematic seminars, workshops, and conferences. More than 20,000 officers of the armed forces of 190 countries have been trained, with courses in English, French, Spanish, Arabic, Russian, and Chinese.

In addition to this overall reference to CPP, in 1984, the Institute organized a Symposium to commemorate the thirtieth Anniversary of The Hague Convention of 1954 and, in 1986, a workshop on "The adaptation of international law on the protection of cultural property to technical developments in relation to modern means of warfare".

Faithful to this tradition, in 2009 the Institute decided to contribute to the call to action promoted by UNESCO by organizing a seminar to celebrate the 10th anniversary of the Second Protocol to the Hague Convention, aiming to spread knowledge on international regulations governing the Protection of Cultural Property in the event of armed conflict, to share information, to facilitate the exchange of knowledge and promote the dissemination among civilian and military actors of the rules governing the protection of cultural property in armed conflicts.

More recently, on the specific request of the same UNESCO, the Institute developed a military manual on the protection of cultural property, with the general objective of further contributing to more tangible dissemination and operational application of the principles and rules governing this field in international and non-international armed conflict. The Manual was the outcome of a series of high-level meetings among international experts (both military and civilians) held in Sanremo between 2015 and 2016.

The Manual[8] combines a military-focused analysis of the existing provisions in the field, with suggestions of best practices at the different levels of command and during the different stages of military operations conducted by land, sea, and air. The text was presented at the concluding conference of the NATO Science for Peace and Security project: "Best Practices for Cultural Property Protection in NATO-led Military Operations" held at the Institute in Sanremo in December 2016.

More recently, another key commitment of the Sanremo Institute in the field of CPP is its contribution to the development of the International Master Programme on "Cultural Property Protection in Crisis Response". The first edition was launched in 2018, promoted by the University of Turin and the University School of Strategic Sciences, in collaboration with the Italian Army, the Carabinieri Command for the Protection of Cultural Heritage Unit, and other relevant institutions and organizations active in the field of CPP and cultural heritage protection.

This program aims to prepare future generations of civil and military professionals by providing them with a multidisciplinary background and advanced expertise in the protection of cultural property in situations of armed conflict, natural disasters and, more widely, in the field of cultural property exploitation as a key element of social and economic reconstruction in affected areas and communities.

---

[8] UNESCO (2016), is available online at the following link: https://iihl.org/wp-content/uploads/2018/01/Military-Manual-EN-FINALE_17NOV-1.pdf.

The importance of the role of cultural property in crisis and post-crisis scenarios has increased considerably in recent years, due to the emergence of new forms of conflicts and the intensification of disasters. It is also for this reason that the Institute, besides regularly renewing its training sessions in this field, during the last years has been keeping a very close contact with the most important institutions in the field of cultural property protection, not only to innovate capacity-building on this topic but also to find new tools and instruments able to multiply the impact of the dissemination initiatives conducted by different stakeholders on the topic.

References

Benvenuti P, Sapienza R (2007) La tutela internazionale dei beni culturali nei conflitti armati, Milan

Caracciolo I, Montuoro U (eds) (2018) CASD, Preserving cultural heritage and national identities for international peace and security, Turin

Convention for the protection of cultural property in the event of armed conflict, The Hague, 14 May 1954, Art. 3

Frigo M (1986) La protezione dei beni culturali nel diritto internazionale, Milan

Geneva Convention for the amelioration of the condition of the wounded and sick in armed forces in the field of 12 August 1949, Geneva (1949a)

Geneva Convention for the amelioration of the condition of wounded, sick and shipwrecked members of armed forces at sea of 12 August 1949, Geneva (1949b)

Geneva Convention relative to the treatment of prisoners of war of 12 August 1949, Geneva (1949c)

Geneva Convention relative to the protection of civilian persons in time of war of 12 August 1949, Geneva (1949d)

GIOIA A (2000) La protezione dei beni culturali nei conflitti armati. In: Francioni F, del Vecchio A, de Caterini P (eds) La protezione internazionale del patrimonio culturale: interessi nazionali e difesa del patrimonio comune della cultura, Milan

Greppi E (2012) I crimini internazionali dell'individuo, Turin

Greppi E (2020) La protezione dei beni culturali nei conflitti armati. In: Rassegna dell'Arma dei Carabinieri, Anno LXVIII – Gennaio/Marzo

International Institute of Humanitarian Law (1986), La protezione internazionale dei beni culturali, Firenze

Kolb R, Gaggioli G (2013) Research handbook on human rights and humanitarian law, Cheltenham-Northampton

Nahlik (1967) La protection internationale des biens culturels en cas de conflit armé. In: Recueil des Cours de l'Académie de Droit International de la Haye

O'Keefe R (2006) The protection of cultural heritage in armed conflicts, Cambridge

Panzera AF (1993) La tutela internazionale dei beni culturali in tempo di guerra, Turin

Toman J (1984) La protection des biens culturels dans les conflits armés internationaux: cadre juridique et institutionnel. In: Etudes et essais sur les droit international humanitaire et sur les principes de la Croix-Rouge en l'honneur de Jean Pictet, Genève-La Haye

UNESCO (2016) Protection of cultural property. Military manual. IIHL, Sanremo

Zagato L (2007) La protezione dei beni culturali in caso di conflitto armato all'alba del secondo Protocollo 1999, Torino

Suggestions for Field Work

Heritage, Development, Community: Methodological Reflections on Research-Action Experiences

Mirella Loda⊙

Abstract This paper discusses the need for interventions by heritage protection experts to interrelate with the social and economic dynamics of the local society. In so doing, the paper stresses the urgency for approaches fit for addressing the complexity of the continuously evolving local systems, and for ensuring community empowerment. Also examined is the effectiveness of technical interventions that are preceded and accompanied by extensive socio-anthropological research into the specific context. Examples are given that show the variety of methods that can be used for this kind of research, and which also proved effective tools for participatory practices.

Keywords Cultural heritage · Socio-anthropological research · Community empowerment

The contributions presented in this volume are the result of experiences of cultural heritage preservation, carried out in vastly different technical spheres, and highly diverse geographical and temporal contexts. They range from interventions for heritage preservation in its physical form, to archaeological research, to museum reconstruction, to a structural analysis of the cultural landscape, to the formulation of protection plans in the context of broader territorial transformation processes.

Despite the diversity of these contexts, the contributions all offer a cue for general reflections on the relationship between development cooperation action for the protection and enhancement of cultural heritage on the one hand, and processes of territorial transformation and socio-economic development on the other.

M. Loda (✉)
Department of History, Archaeology, Geography, Fine and Performing Arts (SAGAS), Laboratory for Social Geography (LaGeS), University of Florence, via San Gallo 10, 50129 Firenze, Italy
e-mail: mirella.loda@unifi.it

M. Loda and P. Abenante (eds.), *Cultural Heritage and Development in Fragile Contexts*, Research for Development, https://doi.org/10.1007/978-3-031-54816-1_21

1 Addressing Complexity

The first point to stress is that each contribution brings us back to what we might call a central challenge with regards to what will be the results of development cooperation actions, namely the in-depth understanding of the local context and the capacity to address its complexity.

This is in fact the precondition for any effective connection between technical intervention and the social dynamics and transformative processes taking place in the area. Taking into account the complexity of territorial systems, deciphering an area's societal articulation and long-term transformative processes provides knowledge of that area, necessary in assessing interventions in terms of their effectiveness and sustainability. This is particularly true for action aimed at safeguarding cultural heritage, which, as a highly symbolic expression of the identity and memory of local communities—and, in the case of world heritage, of all humanity—offers fertile ground for the activation of valid development paths. This is even more the case in fragile contexts. Specific knowledge of the context, of its physical-natural but also social subtleties, and of the evolutionary dynamics at play, is crucial for the realization of a vision, in which the preservation of cultural heritage becomes key for the self-recognition of local communities while functioning as an effective lever of social consolidation and economic advancement.

An in-depth geo-anthropological and social analysis should therefore lay the groundwork for and accompany interventions for the safeguarding of cultural heritage. Only a solid knowledge of the local reality, particularly from a social perspective, allows for an understanding of both the symbolic and concrete meanings that the protective action takes on for local communities, today and in future. It allows for the drawing up of scenarios aimed at the development and activation of endogenous resources—material, economic, human—which heritage intervention could act as a catalyst for. This is because a cultural heritage protection that does not turn its gaze to the future value of the property is inconceivable.

The need for a high-quality cognitive framework brings about the need for highly skilled personnel with adequate disciplinary training and deep knowledge, especially in the field of geo-anthropological and social sciences. In short, the practice of conservation is not a matter for conservation experts alone and cannot be understood solely through a historical lens.[1]

Similarly, it is necessary to come up with an adequate timeframe. Work continuity for scholars and cooperative practitioners should be ensured to the greatest extent possible so as to facilitate the consolidation of knowledge and experience of the context as well as interactions with the local community.

The contributions presented in this volume, offer up evidence for the importance of devoting full, long-term and unwavering attention to the study of the social context as a guarantee for the effectiveness of the intervention.

---

[1] The tendency to conceive heritage exclusively from the technical point of view of archaeology has been recently criticized by Falser and Juneja (2013).

2 Empirical Research and Participative Practices

The second point to stress is that in-depth understanding of the context and sustainable heritage safeguarding can only go hand-in-hand with community involvement and empowerment. There cannot be a universal heritage site that is not first and foremost recognized by the local community.

This approach to heritage protection has now been almost universally accepted on the theoretical level, as confirmed by the contributions to the recent UNESCO Conference *Cultural Heritage in the 21st Century.*[2] More dubious, however, appears to be how it is put into practice in development cooperation projects.

Placing conservation interventions in their contemporary environment means conceiving the local community as a strategic actor in the process, rather than simply providing basic forms of community involvement, which often result in a mere rhetorical addition to otherwise purely technical projects.

So as to really be a means towards general social development, the safeguarding of cultural heritage should proceed through the strengthening of institutions and grassroots democracy. In this paradigm, it is crucial to enact participative practices able to facilitate a deeper understanding of the place, its internal dynamics and workings, and finally, processes of virtuous change. Therefore, the way in which the participative processes are formulated is of paramount importance.

The first step should be to engage in dialogue with all extant institutions directly or indirectly involved both in the safeguarding of cultural heritage and in the urban planning sector, so as to guarantee that heritage preservation policies do not remain isolated sectoral interventions, but part and parcel of the general territorial planning tools of the protected territory, as stated by the ICOMOS *Paris Declaration on heritage as a driver of development* (ICOMOS 2011) as well as by other documents from UNESCO or its advisory bodies (see Introduction).

With regards to institutional personnel, its involvement in extended training activities has proved to be especially effective, such as those provided by Florence University for technical staff from the MoIC-Afghan Ministry of Information and Culture, the MUDH-Ministry of Urban Development and Housing and the Bamiyan Municipality during the preparation of the *Bamiyan Strategic Master Plan* (LaGeS 2018).[3] This solution provided a suitable space–time context in which researchers and practitioners were able to develop a common approach and language for the safeguarding of cultural heritage, within the broader framework of the spatial and urban management of the valley, and to deal jointly with complex methodological and technical aspects.

---

[2] The UNESCO Conference *Cultural Heritage in the 21st Century* (UNESCO 2023) took place in Naples, 27–29 November 2023. One of the six sessions of the conference was entirely dedicated to this topic (Session 4, *Promoting community development through heritage preservation and safeguarding*). However, the contributions of the other sessions abided by this philosophy too.

[3] Thanks to scholarships of the Italian Ministry of Foreign Affairs and the AICS-Italian Agency for Development Cooperation, between 2013 and 2019 eighty people (eleven of them being women) could participate in the Master programs in "Geography, Spatial Management, Heritage for International Cooperation" and in "Urban Analysis and Management" at Florence University.

Interesting results are also to be expected from calling on local researchers to contribute to preparing the knowledge base. Their contribution can be relevant in setting up the research design and identifying sensitive topics. The development of inter-cooperation partnerships with local universities is therefore highly desirable.

The direct and continuous involvement of local staff, researchers, and students has proved extremely effective in developing methodological approaches suited to the local context and for achieving otherwise unimaginable levels of in-depth context analysis. At the same time, it ensures complete community ownership of the process from the very beginning of the interventions.

Local community involvement should also address the wider public of stakeholders and society at large. This involvement can take various forms, the most typical being meetings to allow for an exchange of opinions on the results of the investigations previously conducted on specific aspects of local life, or to illustrate and discuss possible scenarios for the development and strategies connected to the valorization of its cultural heritage. The link between safeguarding cultural heritage and the potential for developing the tourist industry is a major topic to be discussed with local society. It is equally important to publicly address the consequences that prioritizing safeguarding interventions can have in terms of strict rules for other sectors and/or activities. For example, by heightening buildings or developing road networks, we might hope to prevent negative reactions and opposition.

Among the ways by which to involve the wider public, special attention should be devoted to participative practices set up to give voice to underrepresented components of the local population. Depending on the context, this may refer to gender, to ethnic or religious belonging or to other dimensions of the social setting.

As documented by the contributions that make up this volume, geo-anthropological empirical research and fieldwork play an important role in this regard, as an opportunity for direct contact with the population during the production of the knowledge foundations. Firstly, quantitative surveys make it possible to update and integrate the information obtainable from secondary literature (historical, geo-anthropological, sociological, and economic) and the (normally) available statistics, that for these fragile regions are usually highly limited and gathered by international agencies at a rather aggregated territorial level, being therefore of little benefit for the analysis of specific areas or communities.

Secondly, geo-anthropological empirical research can help fill knowledge gaps on various aspects of local life, including the socio-economic conditions of households, the quality of housing stock and housing styles, the status of infrastructure and services, mobility patterns and behaviors, as well as the structure and setup of the productive sectors and so on. These aspects, even if seemingly unrelated to heritage preservation, provide precious information on the material environment surrounding heritage.

Thirdly, through qualitative research techniques such as participant observation, semi-structured interviews or focus groups, geo-anthropological empirical research helps open up a dialogue with the local community on relevant topics such as people's living practices, forms of socialization, how urban spaces are enjoyed, handicraft production modes (especially women's) and so forth. These aspects allow for an in-depth understanding of the socio-cultural environment.

Particularly promising is the use of audio-visual techniques for analyzing changes affecting the cultural landscape, an area as crucial to cultural heritage as it is intricate to analyze. As we will see in the next paragraph, audio-visual techniques prove effective both for analyzing territorial transformations and erosive processes across the cultural landscape and for communicating them to the broader public.[4]

Finally, qualitative empirical research sets the ideal conditions for addressing sensitive issues such as migration processes, settlement dynamics in informal contexts, gender, ethnic belonging, or religious affiliation; all aspects that shed light on the inherent makeup of the local community. Researchers and cooperation practitioners tend to think of the local community as a whole, as a homogeneous entity, while in reality it is made up of various subgroups with often diverging interests. We might even speak of local communities in the plural, instead of as a single community.

Only awareness about the makeup/structure of the local community make it possible to correctly address sensitive issues and mitigate the conflicts inevitably accompanying decisions regarding the use of resources (notably land, financial resources, fundings).

3   Sensitive Issues

The contributions in this volume provide significant examples of how empirical research and in-depth interactions with different components of the local community makes it possible to address sensitive issues and mitigate conflicts. We will briefly reflect on these examples with reference to the topics of gender, interethnic conflicts, and cultural landscape.

As far as gender is concerned, surveys and studies carried out have made it possible to more effectively document the significance that domestic space takes on in Bamiyan and in other rural areas in Afghanistan, as well as in similar realities dealt with in this book. In these contexts, the domestic sphere is relevant not only as a private and family space. Instead, it is extended to encompass parental/clan networks that are sometimes very extensive, and which therefore transcend the strict family perimeter. The extended dimension of families functions as a kind of public (political) sphere and the domestic space works as a quasi-public space.[5] Considerations, which result from preliminary investigations of women's mobility, housing styles, family living practices, and so on, suggest the need for a certain paradigm shift in the way urban public spaces are conceived and designed in contexts of substantial

---

[4] For a more detailed description of the techniques used, see (*Bamiyan Living Culture* 2017 and 2022; Studio Azzurro 2023).

[5] The description of family structure and the role played in it by the female population, as offered by the classic text of Veronica Doubleday, remains instructive and, in many aspects, still relevant (Doubleday 1990).

gender segregation, such as that seen across many regions considered in this volume. In these cases, classically understood open public spaces, like parks, are in fact only enjoyed by the male population. On the other hand, the improvement and enhancement of domestic spaces can strengthen the socialization opportunities and "public" agency of the female population, contrary to what one might assume in the West. Measures in this direction should therefore be included among the primary objectives of urban rehabilitation interventions. Similarly, recognizing the dual significance of shrines and other pilgrimage sites in women's practices—that serve as both spaces for prayer and as open spaces independently usable by women—may open up land-use planning possibilities, aimed at solutions that could improve the quality of their urban experience. In the specific case of Bamiyan, recognizing the shrines' highly symbolic value for the local population (in particular women), ultimately allows those places—although not listed as heritage properties—to be identified as potential objects of protection and enhancement, from a cultural heritage preservation perspective that aims first and foremost at recognition of and enjoyment by the local population.

Moving on to the delicate matter of ethnicity, this is an element of social organization that has received increased political attention in recent decades, and thus also in the field of cooperation.

This attention reflects the importance that this topic has acquired in academia since the 1960s. However, the way ethnicity is often conceptualized entails certain risks. The most common approach sees ethnicity as a primordial condition of the individual. The ethnic group is defined as a unified and stable group of individuals who share an original cultural condition and thus has agency. On a theoretical level, such an ontological reading effectively reifies ethnicity as an innate condition, not susceptible to change. On a practical level, this understanding may lead to a definition of ethnicity which, to the outside observer, explains all societal tensions and conflicts. Consequently, this approach entails an attitude of powerlessness combined with foreignness in the face of tension, that can severely hinder cooperative action.

The opposing perspective refers instead to a reading of ethnicity as a category that takes on relevance only when it "is deployed as a discursive resource for political or economic goals" (Brubaker et al. 2006). Ethnicity—(re)produced in response to the discursive political practice that solicits it—becomes a device capable of hindering the proper interpretation and resolution of tensions and conflicts, and indeed of concealing their underlying causes.

An example might be found in the tension that exists between the Tajiks and the Hazaras, concerning the regional administration's efforts at protecting the cultural landscape in front of Bamiyan's Buddha niches, in accordance with UNESCO's guidance (see Chap. 2.2). The reason behind their different reactions serves as a strong reminder that we should refrain from absolutizing the ethnically grounded viewpoints present in the local context but should instead read them in connection to the political discourse that (re)produces them. The contrasting reactions to the administration's efforts to protect the cultural landscape correspond perfectly to the terms that Naysan Adlparvar uses to define the interethnic dynamic in Bamiyan—"Ethnicity is … an intermittent phenomenon during which people become temporarily ethnicised. They are not always 'Hazarah' or 'Tajik', for example, but become so when their ethnicity is called upon" (Adlparvar 2014, p. 12).

In a context where ethnicity has always been thrown about as a discursive resource in the political arena, it was inevitable that the cultural landscape debate would evolve into a politico-ethnical issue. The difficulties that come from the different positions on the topic of heritage protection were thus charged with tension, owing to the growing rigidity of ethnic factors during the Afghan turmoil. The underlying reasons for the conflict between Tajiks and Hazaras over landscape protection measures in the Bamiyan valley emerged through an empirical survey of the economic base of the two groups, which—putting ethnicity aside—brought out their different economic background (agricultural in the first case, commercial in the second) and thus the differing impact of administrative action. The Hazara group was advantaged, and the Tajik group significantly disadvantaged by the landscape protection policies. In-depth studies have made it possible to move beyond an ethnic reading of the conflict and shift focus to the need to prepare and secure compensatory measures for the disadvantaged farmers.

In the interplay between efforts to protect the landscape and efforts to protect economic interests, the ethnic dimension played a complex and misleading role, stirring mutual resentment and making it harder to strike a balance between the different interest groups. Thanks to the surveys carried out, it was possible to mitigate the conflict between the groups and pave the way for a shared approach to the protection of the cultural landscape.

Finally, it should be remembered that the cultural landscape can represent at the same time an aspect of cultural heritage to be preserved in itself (as in the case of Bamiyan, where the outstanding value of the cultural landscape is mentioned in the nomination), or the setting that includes assets of cultural heritage. In either case, cultural landscape is the sphere in which the intersection between local traditions and modernization and the development drive of the local community becomes visible, with all the transformation and erosion that this may have involved. This makes it especially challenging to look at cultural landscape from a safeguarding perspective.

However, the contributions in the volume demonstrate how an in-depth under-standing of the local culture can pave the way towards approaches to safeguarding cultural landscapes that suit the local context.

They have first shown the urgency of efforts to break out of an abstract logic of pure preservation, and to reinterpret the objective of landscape protection within a strategic vision of spatial organization, informed by the sustainable use and enhancement of local resources.

Moreover, it has become clear that establishing the system of rules with regard to land use and the permissible interventions on the physical landscape in the context of inevitable territorial transformation, fundamental as it may be, does not guar-antee the preservation of the structural elements that make up the cultural landscape. This holds especially in situations-like those handled in this volume—where the concept of landscape is not rooted in the local culture but has been introduced quite recently alongside that of heritage by the international safeguarding agencies. The lack of specific terms with which to translate the concept of landscape into the local language—as is the case of Dari and Pashtu—should be considered an indicator of the difficulties that the local community has in grasping the importance of its preservation as an expression of the distinct, settled relationship between the social community and the territory on which it stands.

Rather, the system of rules for the safeguarding of the cultural landscape needs to be anchored in the widespread awareness of and emotional attachment to the values that the cultural landscape conveys.

Therefore, in cases where the concept of cultural landscape is not rooted in the local culture, the starting point for its preservation should be in interventions aimed at creating a widespread awareness of its relevance both as an emotional connection to the place and as a potential asset for a growth in tourism. It would be appropriate to begin by focusing on projects that advocate a perception and understanding of the landscape as a prerequisite for its protection. An interesting case of this kind is the ongoing documenting and training program set up in Bamiyan. It focuses on audio-visual recordings of characteristic features of the cultural landscape and of the way they are (re)produced or modified by the daily practices of the inhabitants.[6] The aim of the program is to put together an interactive video installation on the subject, open to the wider public, particularly students and schoolchildren of the region, as a sort of education program to the values of the cultural landscape, but also for visitors, in efforts to support mindful and sustainable tourism.

In conclusion, the experiences reported in this volume confirm that in order to be effective and sustainable, interventions for the preservation of cultural heritage should work hand in hand with the social sciences, conceiving safeguarding as part of a general process of social advancement and economic development, of which the local communities ought to be the leading actors and supporters.

References

Adlparvar N (2014) When glass breaks, it becomes sharper. De-constructing ethnicity in the Bamiyan Valley, Afghanistan. PhD thesis, University of Sussex, Institute of Development Studies. http://sro.sussex.ac.uk. Accessed 21 July 2020

Bamiyan Living Culture (2017, 2022) Video installation organized by LaGeS and Studio Azzurro. University of Florence

Brubaker R, Feischmidt M, Fox J, Grancea L (2006) Nationalist politics and everyday ethnicity in a Transylvanian town. Princeton University Press, Princeton

Doubleday V (1990) Three women of Herat. University of Texas Press, Austin

Falser M, Juneja M (2013) 'Archaeologizing' heritage and transcultural entanglements: an introduction. In: Falser M, Juneja M (eds) 'Archaeologizing' heritage? Transcultural research, Heidelberg studies on Asia and Europe in a global context. Springer, Berlin, Heidelberg

ICOMOS (2011) XVIIème Assemblé Générale. In: The Paris declaration on heritage as a driver of development. https://www.icomos.org/images/DOCUMENTS/Charters/GA2011_Declaration_de_Paris_EN_20120109.pdf. Accessed 19 Sept 2020

---

[6] A partial preview of this work was presented at the Cultural Heritage in Fragile Contexts conference, which inspired this volume (Adlparvar et al. 2014).

LaGeS (Laboratorio di Geografia Sociale, Università degli Studi Firenze) (2018) Bamiyan Strategic Master Plan, Polistampa, Firenze

Studio Azzurro (2023) Portatori di storia. Portatori di storie. Mimesis Edizioni, Sesto San Giovanni

UNESCO (2023) Conference. In: Cultural heritage in the 21st century, Naples, 27–29 November 2023. https://ich.unesco.org/en/events/naples-conference-on-cultural-heritage-in-the-21st-cen tury-00977

Heritage and Cultural Activities for Sustainable Development. A Proposal for a Theoretical and Methodological Approach

Emilio Cabasino

Abstract Initiatives organized to promote cultural heritage interventions, and culture in the context of development cooperation more generally, especially in fragile contexts, are always conditioned by the scarce resources available, which are at risk of being allocated to heterogeneous sectors. It is for this reason that such initiatives ought to be conceived, designed and implemented on the basis of a sound and well-defined theoretical and methodological approach. Indeed, in the following pages, we will endeavour to outline such an approach.

Keywords Cultural heritage · Development cooperation · Sustainability

1 Demarcating the Intervention Field

A primary and central issue to be addressed concerns the choice that is to be made as to whether it is appropriate to intervene on cultural heritage in cooperation actions in areas that are not of a high priority, at least not in terms of meeting basic livelihood needs. Indeed, one wonders: can we devote to cultural heritage resources that could be used to provide humanitarian assistance, and that could be allocated to healthcare and to assisting people with disabilities, that could be invested in education, or go towards helping to combat climate change or improve agriculture?

Such a choice is not straightforward and comes with its fair share of ambiguity, and thus should be contextualized, taking two main points into account:

a. The need to move past the emergency stage, where human lives are in danger and the necessary infrastructure for survival is at stake;
b. The need to assign value to the cultural heritage for which there are plans to intervene with regards to: (1) its source community and (2) the assessment method

E. Cabasino (✉)
Italian Agency for Development Cooperation, Via Cantalupo in Sabina 29, 00191 Rome, Italy
e-mail: emilio.cabasino@aics.gov.it

© The Author(s) 2024
M. Loda and P. Abenante (eds.), *Cultural Heritage and Development in Fragile Contexts*, Research for Development, https://doi.org/10.1007/978-3-031-54816-1_22

that is used internationally, relative to the classifications as well as the historical framework that generated it.

Therefore, it seems wise for development cooperation interventions in the cultural sphere to be designed and implemented only after those connected to emergency and humanitarian aid, which are aimed at people's survival.

On the other hand, the importance of the broadly understood cultural dimension of individual well-being, even in precarious living conditions and distressing contexts, should not be underestimated. Cultural baggage and identity are intrinsic to human beings, embedded in an organized social environment, enveloping and conditioning them even while being physically removed from the places they usually reside in.

Interventions within the cultural sphere can thus also be implemented in contexts of humanitarian aid (e.g., in favour of communities displaced from their original residences), just as they can involve physical/material or intangible goods, identified primarily in accordance with the classifications recognized by the international Conventions that concern them (predominantly those of UNESCO[1]).

Material and physical cultural heritage is the most obvious and recognizable (archaeological sites, monuments, the museums and artworks contained within them, libraries, archives, etc.) and most often linked to its touristic potential. No less important, however, is that which is encompassed by the 2003 UNESCO Convention, defined as intangible cultural heritage. The customs, rituals, traditions, ceremonies, manufacturing/crafts, land use or relationship with the land, as well as traditional community practices, all of which help define its identity and foster cohesion.

Lastly, in the context of development aid, sectors of the so-called cultural and creative industries ought to be considered as well (publishing, audio-visual and cinema, design and fashion, crafts, etc.). The latter were indeed the subject of intervention for the 2005 UNESCO *Convention on the Protection and Promotion of the Diversity of Cultural Expressions*.

---

[1] In particular, the UNESCO *Convention on the Protection of World Cultural and Natural Heritage* of 1972; the 2003 UNESCO *Convention for the Safeguarding of the Intangible Cultural Heritage*; the 2005 UNESCO *Convention on the Protection and Promotion of the Diversity of Cultural Expressions*; the 1970 UNESCO *Convention on the Means of Prohibiting and Preventing the Illicit Import, Export and Transfer of Ownership of Cultural Property*; and the 1954 UNESCO *Convention for the Protection of Cultural Property in the Event of Armed Conflict*. An interesting approach is also that of the 1954 *European Cultural Convention* and the 2005 *Faro Convention, on the Value of Cultural Heritage for Society*, which emphasizes the importance of cultural heritage in connection to human rights and democracy and promotes a broader understanding of cultural heritage and its relationship with the communities and society at large.

2   The Sector's Disciplines and Reference Bodies

In this contribution, we limit ourselves to discussing the disciplinary fields and theo-retical approaches considered most useful in defining the forms of intervention and the suitability of public resources allocated to support cultural heritage, and culture more generally, addressing them as factors of growth and development.

First and foremost is the approach that fits into the well-defined set of regulations, forms, and procedures for the analysis and monitoring connected to the UNESCO Conventions mentioned above. These put together an in-depth definition of the theo-retical framework for interventions and operations, from design to implementation, as well as for the monitoring and the assessment of the interventions.[2]

Next, the most important area of study and research is undoubtedly that of the Economics of Culture, a discipline around which a vast and varied debate has been developing since the 1970s. It has produced a number of national and international studies that have investigated and illustrated the role that culture can play in the economic growth of contemporary society, as well as its very nature and the suitability of the public resources that support it. It is not possible here to go over the arguments and differing positions that characterize this disciplinary field, but it is important to mention that the methodological tools for measuring the economic dimension of culture are now well-established quantitatively and, where possible, qualitatively.

In this endeavour, the European Statistical System of the European Commis-sion has played an important role. In 2012, for example, it published the "Final Report", titled *CULTURE European Statistical System Network on Culture* (Euro-pean Commission (2012, 2019), which puts forward an interesting up-to-date summary of cultural activities (Fig. 1).

A methodological framework of great interest for interventions relating to Devel-opment Cooperation is certainly the one put forth by UNESCO, referring to the 17 goals set in the *2030 Agenda for Sustainable Development* (and relevant targets). The UNESCO framework identifies 22 cultural statistics variables and groups them into 4 different thematic indicators, aiming to measure and monitor the sector's economic activities (UNESCO 2019) (Fig. 2).

Approaching the cultural sector as a feature of development, one should also point out the work carried out by the OECD on determining tools for analysing and designing cultural policies (OCSE-OECD 2022).

Research and studies associated with development cooperation interventions in the cultural sector can also in a way be associated with the economics of happiness, which investigates the factors that engender human well-being. Among such factors, participation in the cultural practices of the community is one that is seen as extremely important for those who are part of it (Nikolova and Graham 2020).

---

[2] See https://whc.unesco.org/en/culture2030indicators/#publication. Also of great interest from a UNESCO development cooperation perspective, are all the tools and the bibliography relevant to development, and the role of Intangible Cultural Heritage; see the Kit of the 2003 UNESCO *Convention for the Safeguarding of the Intangible Cultural Heritage* (UNESCO 2009–2019).

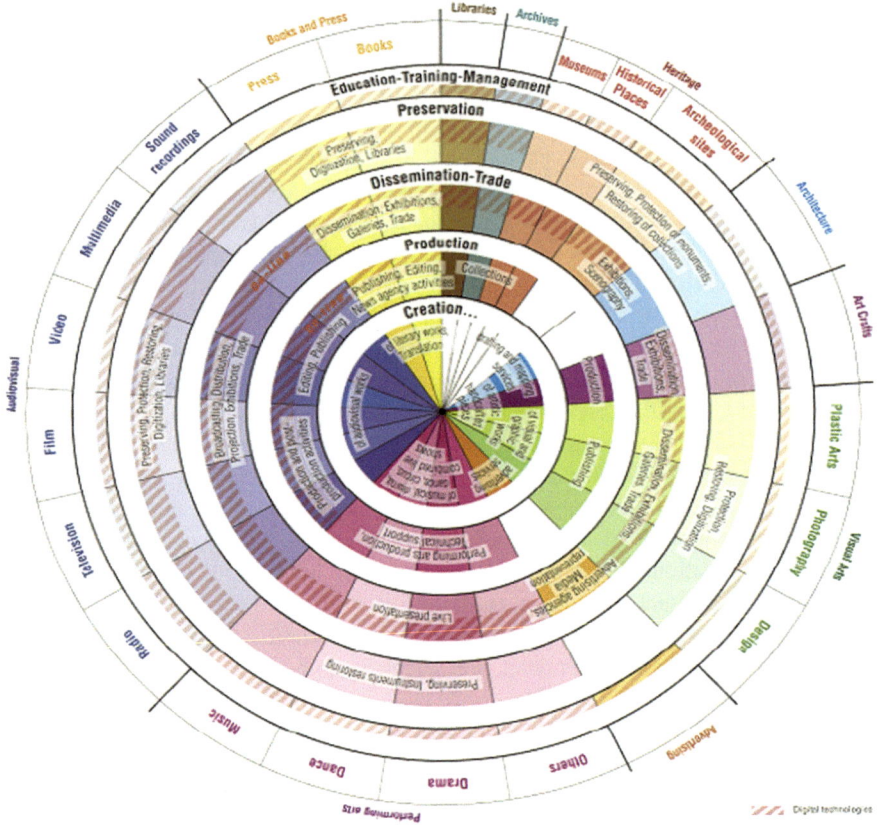

N.B.: *This visual matrix of culture is of course a digest representation whose aim is to give a general view of the system. It is, however, not possible to list all the cultural activities in that layout. For a detailed list of cultural activities, further tables are presented.*

Fig. 1 ESSnet-culture framework: cultural activities (Fig. 4, p. 48 Final Report, 2012)

In this regard, mention should also be made of the attention given to this issue in the United Nations system, which since 2012 has established the position of Special Rapporteur in the field of cultural rights. The task of such Rapporteur is to identify best practices in the promotion and protection of cultural rights at the local, national, regional and international levels and to identify obstacles to the promotion and protection of cultural rights, and make recommendations to the UN Council on ways to overcome them (United Nations 2023).

Moreover, it is important to mention the attention paid to the cultural context within the activities and programmatic guidelines presented in 2015 in the *Sendai Framework for Disaster Risk Reduction 2015–2030*, as part of the United Nations *Office for Disaster Risk Reduction* exercise (UNDRR 2022).

Fig. 2 UNESCO—thematic indicators for culture in the 2030 agenda

Finally, the importance of the cultural aspect both from the perspective of both economic as well as individual and collective well-being presented in Kate Raworth's 2017 ground-breaking study on sustainable development should be mentioned as relevant. The graphic illustration of development as a "doughnut," rather than as a GDP vector graph, highlights with great clarity the importance of a "social base" (and, therefore, social justice) for truly sustainable development to be accomplished.[3]

3 From "Cultural Cooperation" to "Development Cooperation in the Cultural Sector"

As mentioned in the previous paragraphs, the most straightforward, as well as controversial, area of intervention in this sector is "tangible" cultural heritage (archaeological, artistic, monumental, bibliographic, archives, etc.), which represents the material legacy of the past and evokes instantly recognizable and vivid memories for its people.

The material cultural identity of a community: in this case, the need to intervene should not be questioned, as long as the intervention comes after ensuring survival and after rescuing lives. Development cooperation should, in this case, be based on the recognition, preservation and enhancement of the community's founding values,

---

[3] See specifically, Ch. 3, which discusses "Nurture Human Nature" (Raworth 2017).

while also aiming at creating activities and services that can generate profit for the community and, therefore, foster sustainable development.

It should be stressed here that initiatives of this kind, which are supported by the Italian Agency for Development Cooperation (AICS), are often the result of extensive research and study missions that Italian universities have been embarking on with admirable dedication and excellent results. Said initiatives are indeed recognized within the academic world and are supported by resources made available by the Ministry of Foreign Affairs and International Cooperation (MAECI) through the Directorate General responsible for cultural cooperation.

Therefore, the difference between "cultural cooperation" and "development cooperation in the cultural sector" lies precisely in the purpose of the two activities, which are complementary but also distinct. The former is connected to academic research (historical, philological, archaeological, artistic, monumental, anthropological, etc.) and to the interchange of professional expertise among the countries being studied. The latter is aimed at activities carried out in the cultural sphere (in its broad sense) that can contribute to the development (again, in its broad sense) of the country in which they are carried out.

As for potentially development-generating activities, some examples include:

1. Direct interventions on heritage

   a. Artefact conservation and restoration
   b. Cataloguing, surveying, documenting (drawings, photos, videos, etc.)
   c. Interventions connected to site accessibility and enjoyment of the visit

2. Other services connected to Cultural Tourism

   a. Ticketing and reservations
   b. Tour guides
   c. Feature publications
   d. Transportation
   e. Hotel accommodation/catering

3. Cultural and creative industries

   a. Artistic handicrafts
   b. Video products
   c. Fashion and design relating products
   d. Events

To conclude this paper, an attempt will be made to outline an initial list of features and factors that should be taken into consideration when setting up and implementing an intervention project aimed at achieving development by supporting the heritage and cultural activities sector.

4   An Initial Outline of the Approach to be Followed in Cooperation Initiatives or Programmes Involving Heritage and/or Cultural Activities

Following the procedures of the Italian Agency for Development Cooperation (AICS), an intervention in cultural heritage and/or culture more generally, should, broadly speaking:

a. be relevant to one or more countries among those identified by the Ministry of Foreign Affairs and International Cooperation (Italian Ministry for Foreign Affairs 2023);

b. keep in mind that in planning the interventions, the Italian Cooperation focuses on the 5 "P" of the UN *2030 Agenda* (People, Planet, Prosperity, Peace, Partnership), aims at achieving one of the 17 goals, and maintains as its point of reference the targets and implementation tools that are associated with each goal;

c. be agreed upon with the relevant Italian Embassy and negotiated in advance or, better, by answering to a specific request that is sent by local government authorities.

More specifically:

d. respond to a development demand while respecting local identities and the needs of the community in question;

e. identify, in the most accurate way possible, the scope(s) of the intervention within the international classification systems mentioned in the previous paragraphs, and especially that put forth by UNESCO;

f. provide useful tools of determining and gauging baseline, and be able to measure quantitatively as well as (where appropriate and possible) qualitatively, the impact of interventions;

g. strike a balance between the economic-commercial dimension and respect for the preservation and conservation of the cultural assets that are the subject of the interventions;

h. identify the local professionals involved in such initiatives, and ensure that they benefit the most from collaborating with international professionals;

i. make sure that said benefits always go two ways, i.e., that the international professionals working locally act respectfully and learn about the cultural, social, administrative, and procedural practices of the context in which they operate.

If services or business activities are to be carried out, especially of a touristic nature, the intervention must:

j. act with respect towards the social, environmental and cultural environment in which it is implemented, ensuring fair and widespread distribution of any anticipated profits;

k. aim at a gender balance, making sure to involve women;

l. consider the added value that comes with promoting accessibility and a broader enjoyment of the site, paying special attention to people with disabilities;

m. encourage the consumption of local products, as well as community tourism, by involving local people in hospitality.

The points laid out above are by no means exhaustive in determining the approach that should be followed in emergency or development cooperation interventions connected to cultural heritage and culture. We hope, however, that they can help establish an initial checklist useful to guide action, taking into account the many multifaceted variables that should be considered when intervening in this sensitive, complex but wonderful field.

References

European Commission (2012) European statistical system. Culture European statistical system network on culture. https://ec.europa.eu/assets/eac/culture/library/reports/ess-net-report_en.pdf. Accessed 18 Feb 2023

Italian Ministry for Foreign Affairs (2023) International cooperation for development. Three-year planning and policy document 2021–2023. https://www.esteri.it/wp-content/uploads/2021/11/Schema-di-Documento-triennale-2021-2023.pdf. Accessed 5 Jan 2023

Nikolova M, Graham C (2020) The economics of happiness. GLO Discussion Paper, No. 640. Global Labor Organization (GLO), Essen. https://www.econstor.eu/bitstream/10419/223227/1/GLO-DP-0640.pdf. Accessed 18 Dec 2022

OCSE-OECD (2022) The culture fix. Creative people, places and industries, local economic and employment development (LEED). https://www.oecd-ilibrary.org/urban-rural-and-regional-development/the-culture-fix_d493fdac-it

Raworth K (2017) Doughnut economics. Seven ways to think like a 21st-century economist. Random House, London

UNESCO (2009–2019) Kit of the convention for the safeguarding of the intangible cultural heritage. https://ich.unesco.org/en/kit#8. Accessed 19 Dec 2022

UNESCO (2019) Thematic indicators for culture in the 2030 agenda. https://whc.unesco.org/en/culture2030indicators/. Accessed 20 Dec 2022

United Nations (2023) Office of the High Commissioner. Human rights special rapporteur in the field of cultural rights. https://www.ohchr.org/en/special-procedures/sr-cultural-rights. Accessed 25 Jan 2023

United Nations Office for Disaster Risk Reduction (UNDRR) (2022) Sendai framework for disaster risk reduction 2015–2030. https://www.preventionweb.net/publication/sendai-framework-disaster-risk-reduction-2015-2030?_gl=1%2A16ceonc%2A_ga%2AMTU0ODI2MzA2Ny4xNzAxOTYyMjQ3%2A_ga_D8G5WXP6YM%2AMTcwMTk2MjI0Ny4xLjEuMTcwMTk2MjM2Ni4wLjAuMA. Accessed 29 Nov 2022